Achieving Sustainability: The Ultimate Human Challenge

Karen Blincoe

Achieving Sustainability: The Ultimate Human Challenge

Critical Barriers and Future Perspectives

Springer

Karen Blincoe
Dronningmolle, Denmark

ISBN 978-3-031-10022-2 ISBN 978-3-031-10023-9 (eBook)
https://doi.org/10.1007/978-3-031-10023-9

This Springer imprint is published by the registered company Springer Nature Switzerland AG
The registered company address is: Gewerbestrasse 11, 6330 Cham, Switzerland

To David, Sophia, Chloe, Alexander and Jacob

Preface

From the age of 7–13, I used to run away regularly. My parents' relationship was fiery, and arguments would happen from time to time. I would get fed up and leave home, which meant walking out the back door, down through the garden, past the beehives and the vegetable garden, the blackcurrant bushes and cherry trees, into our farmer neighbour's corn field. There I would hide among the wheat or barley (my favourites), lie down and look up at the sky, watch the clouds, the ladybirds, the ants and other creepy-crawlies on the ground, listen to the birds above and the sound of the rustling corn—or I would hide in the long grass and wildflowers in the hedgerows. There were many natural places to hide in, in those days. My parents and siblings did not worry about me disappearing. They knew I would be back when I got hungry, generally for the evening meal.

We lived in the country surrounded by fields, woods and close to the sea. We had a big garden and grew nearly all of our food. So my life as a child was living in, around and with nature. It was part of my DNA and has continued to be so all my life. My love for nature knows no boundaries. I live surrounded by it, and it's a factor in the way my life and career have unfolded and why sustainability has been an essential part of that journey. Another factor that has contributed to my interest and passion for these issues is how we lived during my childhood. We lived an almost sustainable life. By that, I mean that my father, who, in addition to being the village schoolmaster and church singer, was also a beekeeper and managed our large vegetable garden with the help of us children.

The garden covered all our vegetable and fruit needs for the year, such as potatoes, carrots, onions, leeks, Brussels sprouts, peas, cucumbers, beets, kale, cabbages, beans, and celeriac. What could not be kept in the ground through the winter was deep-frozen. My mother made pickled beets and cucumbers. The fruits we grew were strawberries, blackcurrants, redcurrants, gooseberries, raspberries, cherries, apples, and plums. Some were frozen for desserts for the winter months, some were made into jams, and some were turned into cordials or other condiments. We had a cellar full of pickles and preserved foods for the winter, and we had a section in the communal deep freeze, full of vegetables, fruits, and meats. The communal deep freeze was a building consisting of one room covered by a large deep freeze divided into compartments. The citizens in the municipality were each given a compartment where they could store their frozen foods.

Meat was part of my upbringing, always on the menu with potatoes. Early in my childhood, we kept a pig in the garden. It ate all our scraps of food during the year and was slaughtered at Christmas. Every morsel was used: cuts for roasts, chops, mince, a variety of sausages, pates, cuts for stews, black pudding, etc. My father went hunting, and my brothers fished. So, we had all the meat and fish we needed. My mother looked after us, in addition to running the school library. She cooked, cleaned, washed, ironed, sewed, darned, and baked with the help of her daughters, my sister and me. There was no waste, no unnecessary consumption. Presents for birthdays and Christmas were sensible, perhaps with the addition of a doll or book from time to time. New clothes were a rarity. We lived simple lives but rich lives. Rich in meaning, traditions, values, and learning.

These events tell me how important my childhood experiences were in shaping my life's journey. The first was the love for nature. The second was resource responsibility and self-sufficiency. The third and fourth elements that gave depth to the other two were the intellectual and spiritual aspects of my life. School and learning were ingrained elements in our everyday lives. The school was situated at one end of the house, while we lived at the other. My father taught three different years simultaneously and so, storytelling was important when learning about topics such as history and religion. My father was a follower of N. F. S. Grundtvig's ideas, particularly those related to education. Grundtvig (1783–1872) was very influential in Danish society from 1820 onwards. He had many roles as pastor, teacher, poet, philosopher, politician, and historian. He was visionary and radical and applied his ideas to society. Grundtvig's views on education went against the norm of that time. He advocated participative and action learning and learning for life rather than focussing on educating scholars, and he was one of the founders of the folk high school movement. My father's other educational role model was Christen Kold (1816–1870), a teacher and folk high school principal famous for his storytelling abilities. I grew up living in a village school environment with these approaches to education. That had an impact. The ideas of life-long learning as a participative and personal development activity have, as such, been founding pillars in my teaching approach.

My mother ran the school library, where the local community came to borrow books. The library was another place for escape and solitude, shared by most of the family members. I loved the books and read from early age authors such as Jean-Paul Satre, Albert Camus, Simone de Beauvoir, Franz Kafka, etc. Reading became a passion. The spiritual aspect was also an innate part of my early life that has remained with me. Early on, spirituality was related to my parents' beliefs. They were Christians and keen churchgoers. I prayed and had a sense of something or someone called God. But it was not until I discovered a different kind of spirituality at the age of 19 at a folk high school that I found something that I could relate to and feel supported by. A folk high school teacher introduced me to Kahil Gibran (1883–1931), a Lebanese-American poet and writer famous for "The Prophet" (1923). In his books, I found the wisdom, that I have pursued since then through a variety of practices taught by many wonderful spiritual teachers.

The fundamental learning, both practical and theoretical, that was available to me as a child, I am very grateful for and feel enriched by. That early learning has supported me throughout my personal and professional life and is a key to my passion for and relationship with sustainability. However, even though I respect the past, I do not believe that the solution is to go back to that way of life or to that time. However, learning from history is essential to shaping the future.

Dronningmolle, Denmark Karen Blincoe

Acknowledgements

This book is an adaptation and update of the PhD thesis that I completed in 2020. The support and contribution from the Research Departments at The School of Art at Kingston University, London, UK, and The School of Art at Brighton University, UK, where I was a student while I researched and developed the thesis, have been invaluable. My supervisors at both places guided and supported me expertly through the process. Our regular meetings and conversations were essential in helping me to clarify what I was doing and where I was going. At Kingston University, principal supervisor Paul Micklethwaite and second supervisor Christopher Doll kept me focused during the final two years of thesis writing. At Brighton University, principal supervisor Prof. Jonathan Chapman and second supervisor Paul Hopper kept me on the right path during the first years of research.

I owe a special thank you to Prof. Emerita at Kingston University, Anne Boddington, who allowed me to complete the PhD at Kingston University.

I thank the three wonderful and inspiring case story teams: Søren Hermansen, Malene Lunden, and Michael Kristensen from Samsø Energy Academy; Sister Jayanti, Golo Pilz, and Sonja Ohlsson from Brahma Kumaris, and Claus Stig Pedersen, Per Henning Nielsen and Hannele Sørensen from Novozymes A/S. They generously shared their experiences from the change processes, and they are the kind of role models we need and can learn from as we take the steps towards achieving the necessary transition to sustainability.

Many people have inspired me on the long journey from research for the PhD to writing the book. I want to thank my wonderful and wise board colleagues and staff at Chora Connection from 2014 to 2019 in Copenhagen. The work and discussions inspired me to continue the research and complete the PhD, even though it was a stressful time. My gratitude also goes to the many loving friends and colleagues in Denmark and abroad who have supported me both in the external career world and on my internal spiritual path.

The work at the ICIS Centre (2001–2014) would not have been possible without the continued support from the ICIS board, for which a special thank you goes to my supportive sister and great friend, Birte-Marie Jørgensen, to colleague Vibeke Grupe Larsen and colleague and long-time friend, Steinar Valade-Amland. In this context, I am indebted to my daughter, Philippa Shepherd (a sustainability researcher), for generously developing the sustainability content in ICIS, working on the Rødding Manifest and on the Mali project—and was also an ICIS board member for many years. Without her contribution, we would not have achieved what we did.

The experience heading Schumacher College in Devon, UK (2006–2008) was invaluable. It broadened my knowledge of sustainability and showed me how important it is for an organisation to work holistically inside and out. The many conversations with the visionaries who taught at the college at that time inspired my future work. The first time I attended Schumacher College as a participant in 1998, I was inspired to create a similar platform in Denmark, which subsequently became the ICIS Center.

I am grateful to the Publishers, Springer Nature, and especially Margaret Deignan for taking on this book project, even though I am a novice, and for answering all my queries when in doubt about the process.

Equally, I am indebted to Tim Harper, Visiting Professor, Newmark Graduate School of Journalism, City University of New York for helping me through the first months of dealing with publishing information and communication. It was a new field for me to enter, and without Tim's reassurance and comments, I would have hesitated to embark on the book project. I am also grateful to Nancy Bobrowitz, Tim's wife, the best client I have ever had in my design practice in London so many years ago, and who was also remained a valued friend.

Regarding the inner work, there are many special people to acknowledge. I am grateful to my long-time friend Mary Aver, a healer, shaman, and spiritual teacher in London, who created the Wise Women network in Denmark and has been a continuing source of wisdom over many years. The Wise Women network in Denmark lasted 20 years and was joined by extraordinary women, especially Ulla Heegaard, Pia Stærmose, Anne Echwald, and Margit Thomsen who have remained close friends.

My gratitude includes my own and extended family members. A special thank you goes to my daughters, Anna Blincoe, a fine artist who made the front cover illustration for the book, and Philippa Shepherd, whose special contribution I mentioned above, and their families for sharing ideas on sustainability issues, and supporting me during the ups and downs over the years—for their continued generosity, love, and care.

My final debt I owe to Mervyn Kurlansky, my wonderful husband, for continuously supporting me, no matter what, and for encouraging me to do what I do. Paying the bills when I haven't been earning, tolerating when I was somewhere else in my mind, and at times working day and night. Given support when I have been in doubt, read the chapters in the book, given constructive feedback, and helped with practical matters, such as designing the diagrams for the thesis and beautiful front cover ideas for the book.

This book is dedicated to our grandchildren: David Christian Blincoe Michaelsen, Sophia Leonor Shepherd, Chloe Sofie Blincoe Michaelsen, Alexander Graeme Shepherd, and Jacob Kurlansky-Thom. They have intelligence, interest and a keen awareness of what is happening in their world. I believe these five young people have the resilience, courage, and what it takes to meet the challenges that lie ahead and contribute to rebalancing our world.

With love and gratitude to you all.

Contents

About the Author

Dr. Karen Blincoe is an internationally acknowledged expert on sustainability and related fields. For over three decades, she has pioneered sustainability initiatives such as founding the ICIS Center and co-founding Chora Connection in Denmark. She has developed master's programmes in sustainable leadership, headed EU programmes on sustainable design, and lectured and taught on all continents. Her unique position in addressing sustainability lies in her comprehensive understanding of the issues at stake, having gained both academic as well as practical experience in the field. Karen has previously been Head of Department at the Danish Design School, Director of Schumacher College, UK, Visiting Professor at the University of Brighton, and Reader in Sustainable Futures, Goldsmith's University, London. Karen is married, has two daughters and four grandchildren, is based in Denmark, and divides her time between being a writer, lecturer, examiner, and a board member.

Abbreviations

APEC	Asia-Pacific Economic Cooperation
BK	Brahma Kumaris
C2C	Cradle-to-Cradle
CG	Corporate Governance
COP	Conference of the Parties
CSD	Commission on Sustainable Development
CSR	Corporate Social Responsibility
DAVOS	Name of Alpine town in Switzerland
DIIS	Department of Industry, Science, Energy and Resources
DPI	Department of Public Information
DSP	Dominant Social Paradigm
ECOSOC	United Nations Economic and Social Council
EEA	European Environmental Agency
EEE	Environment, Ethics, Economics
ERP	Education for Rural People
ESG	Environmental, Social and (corporate) Governance
EU	European Union
FAO	Food and Agricultural Organization
GDI	Gender-related Development Index
GDP	Gross Domestic Product
GHG	Greenhouse gas
GNH	Gross National Happiness
GNHI	Gross National Happiness Index
GNP	Gross National Product
GPI	Genuine Progress Indicator
HDI	Human Development Index
HPI	Happy Planet Index
ICIS	International Centre for Creativity, Innovation and Sustainability
IIED	International Institute for Environment and Development
IPCC	Intergovernmental Panel on Climate Change
ISO	International Organization for Standardization

ISO14040/44	Set of principles and framework for life cycle assessment
IT	Information technology
IUCN	International Union for Conservation of Nature
LCA	Life-Cycle Analysis or Assessment
NASDAQ	American Stock Exchange
NGO	Non-Governmental Organisation
NRTEE	National Round Table on the Environment and Economy
OECD	Organisation for Economic Co-operation and Development
PPP	People, Planet, Profit
PR	Public Relations
SD	Sustainable Development
SDGs	Sustainable Development Goals
SEA	Samsø Energy Academy
SME	Small and Medium Enterprises
SPI	Social Progress Index
TBL	Triple Bottom Line
TEF	The Ecological Footprint
TES	The Environmental Space
TNS	The Natural Step
UN	United Nations
UNDP	United Nations Development Program
UNEP-SPD	United Nations Environment Program for Sustainable Product Design
UNFCC	United Nations Framework Convention on Climate Change
UNHCR	United Nations High Commissioner for Refugees
UNICEF	United Nations Children's Fund
WBCSD	World Business Council for Sustainable Development
WCED	World Commission on Environment and Development
WCS	Wildlife Conservation Society
WEC	World Economic Forum
WWF	World Wildlife Fund

Chapter 1
Introduction

Abstract This chapter introduces the topic, the goal, and the content of the book. The first part sets the scene. It describes the context and reason for developing a PhD thesis and writing a book on achieving sustainability including personal and professional experiences. It defines the challenges facing us, particularly climate change and other environmental crises, as well as the interventions needed. The approach and the research underpinning the book are explained, including an overview of the broad spectrum of references obtained from academic, personal, and other sources. The second part of the introduction outlines the content of the individual chapters. Each examines a specific aspect of sustainability and its applicability, starting with defining the essence of the construct and the challenges related to the current political and economic paradigm. Profound societal changes in history are examined as well as current practical initiatives and barriers to implementing sustainability.

1.1 The Topic

> Like the caterpillar that wraps itself up in its silken swaddling bands prior to its metamorphosis into a butterfly, we have wrapped ourselves in a tangled skin from which we can emerge only by going through a similarly dramatic transformation (Goodwin 2007).

Sustainability is the topic of this book, especially the challenge of its implementation and achievement. It's a narrative, that unfolds the concept from its launch in 1987 till today seen from the perspective of a professional who has worked with, in and around the sustainability field for more than three decades. It is a perspective that considers sustainability to be the only enlightened model for the future of humanity, and one that is urgently needed now. The perspective is northern European with a global reach.

The reason for undertaking a PhD and subsequently writing this book arose from working with sustainability for many years. I founded and developed the ICIS Centre from 2002–2014 (an educational centre for design and sustainability in Denmark), was Director of Schumacher College from 2006–2008 (an advanced educational centre for sustainable learning in the UK), and co-founded and developed Chora

K. Blincoe, *Achieving Sustainability: The Ultimate Human Challenge*,
https://doi.org/10.1007/978-3-031-10023-9_1

Connection (now Chora 2030) from 2013–2018 (a foundation for experimentation and implementation of sustainability in Denmark).

Sustainability is an all-encompassing construct. Multi-dimensional, multi-disciplinary, transnational, transcultural, and essentially holistic. It is the ultimate model for a balanced world and one that promises to sustain human life for generations to come without destroying the biosphere and ecosystems. As such, it represents the next logical and evolutionary step for the human species. The sustainability concept is relatively new in a historical context even though environmental issues have existed for centuries. The effects of global warming, caused by the industrial revolution began to be observed and felt in the beginning of the twentieth century and were acknowledged by the world governments and the fossil fuel industries in the late 1970's.

Mainstream activities related to implementing sustainability have clearly progressed too slowly to effectively address what is needed. We have not, as yet, been able to halt the negative impacts on our societies caused by unsustainable development rooted in the growth paradigm. The perils have increased as the effects have crossed several planetary boundaries identified as critical thresholds, tipping points, by the Stockholm Resilience Centre. Crossing these boundaries indicate that critical ecological functions, necessary to safeguard humanity's safe operating space, are seriously under threat. Climate change, biosphere integrity, biogeochemical cycles, land system change, and 'green water' (water needed to sustain ecosystems, not just people) have all been transgressed. The planetary boundaries were created by the Stockholm Resilience Centre, an international research centre focusing on the scientific aspects of sustainability and resilience. The boundaries were identified in 2009 by a group of scientists, led by the former head of the resilience centre, Johan Rockstrom. They consisted of nine processes, that regulate and adjust the Earth's systems with regard to the atmosphere, sea and land.

The recent years have seen a plethora of publications predicting an apocalyptic future as a consequence of the adverse impacts of our unsustainable behaviour. The seriousness of the situation cannot be underestimated. The latest IPCC (Intergovernmental Panel on Climate Change) Report, collated from thousands of scientific papers paints a bleak picture of the future (Sect. 2.4).

Arguably, the sustainability agenda, first launched by the United Nations (UN) in 1987, has in the main been advanced by individuals such as researchers, writers, visionaries, pioneers, enlightened entrepreneurs, business leaders, and a few visionary politicians, NGOs, and research institutes, who have seen the damaging effects of our unsustainable behaviour and who have believed in the importance of sustainable development. While governments have played a pivotal role in putting sustainability and environment on the political agenda, they have not acted sufficiently nor expediently. Aside from some ambitious policy and regulatory objectives, the main message has been that it is the individual's responsibility to respond, to change behaviour, rather than the government's responsibility to take the lead and initiative. Research shows that a transition to a state of sustainability may only be possible given sufficient time and through a synergy between a constellation of profound change factors. This can only be achieved with the intervention of the state in a systemic and coherent way through policy, regulatory and infrastructural change. The lack of government-lead course correction measures is a growing concern as without it, business agendas based on economic growth above all else will continue to thwart any headway toward a sustainable transition.

The two key challenges to be met are: (1) to muster the political will and determination to implement sustainability politically, economically and socially, and (2) to mainstream appropriate tools, skills and methods, including technological solutions, necessary to achieve a sustainable development. In order for this to occur, and to understand why it is taking such a painfully and distressingly long time, we need to look at the tangible and intangible barriers to a sustainable transition that exist innately in our societies and in the construct of sustainability. These barriers need to be defined, examined and understood to be able to assess what is required to overcome them in order to achieve the necessary change.

The goal of the book is to examine and provide clarity on these issues. It is an enquiry into the complex field of sustainability, its nature, context, application, and challenges. It is an enquiry into the nature of the barriers and offers an assessment of whether sustainability is even possible. Essentially, the book addresses the questions: Is sustainability an achievable utopian ideal? If yes, what role does the 'vision' of sustainability actually play? In order to discuss these questions, the book delves into and examines historical attempts to radically change societies. Not from the perspective of new conventional development theories but from the perspective of utopian thought, models and experiments, which symbolise and represent the radical nature of sustainability. The book argues that utopian thinking is critical for making transformational change to a system or society as it is only through imagination that we can create a future beyond the confines of our current thinking and context. Using three contemporary case studies, where an organisation, an island and a multinational corporation have engaged in applying strong and radical sustainability principles and methods, this book shows that under the right circumstances, and with profound dedication and will, the utopian vision of sustainability might be realised, simply through a matter of choice and action.

1.2 The Approach

The book will, through the following seven chapters, address different aspects of sustainability. The approach is based on my belief that a solution to a problem can be found in the understanding of the problem itself. The chapters therefore explore the nature of the current problems. They include an account of the essence of sustainability and the background from which it emerged. In addition, they include a documentation of the principles, models and methods that have been developed in an attempt to implement and apply the ideals. The chapters include a discussion of the link to utopian thought as mentioned above, the lessons learnt from practical examples, and the resistance to sustainability that are innate in the current economic, political, and cultural paradigm. Other topics cover the last two years of global change caused by the pandemic as well as sustainability seen from a personal perspective with the purpose of exploring external influences and what we can do individually.

1.3 The Research

The book is based on a PhD thesis that I completed in 2020. I wanted to turn the research into a book, and add my thirty+ years' experience working with sustainability to the academic research I had conducted. The book, therefore, consists of a mixture of sections from the thesis as well as some updated information, and personal points of view. The aim has been to address the topic from a holistic perspective, as I believe this is the only way we can fully understand what is needed.

The field of sustainability is relatively new as an academic discourse and field of study. It is within the past three decades, specifically after the launch of the Brundtland Report in1987, that scientific research became focussed on topics of sustainability and sustainable development. Consequently, the amount of research and reference material was limited until early this century. My research for the PhD started in 2010. Some parts of the thesis were written during the first few years, updated and finalised in 2019 and 2020. I wrote the book between 2021 and 2022. Choosing the reference material presented me with a challenge in 2010 as the amount of relevant and qualitative data was relatively limited. As sustainability covers many fields, and the primary focus early on was on environmental problems, knowledge of the broader perspectives balancing the environmental, social, and economic issues was limited.

The references that interested me had to include a wide variety of aspects, including solution-based theories as well as research that discussed both global, national, cultural as well as personal barriers to sustainable development. I consulted the UN and the IPCC (Intergovernmental Panel on Climate Change) reports frequently, to access the most valid and factual information. The IPCC reports are collations of the scientific work of thousands of researchers worldwide. They are regarded as reliable sources of environmental and sustainable development data. In addition, The Stockholm Resilience Centre, a highly respected research centre focusing on global sustainability and the nine planetary boundaries, was therefore also an essential source of knowledge and information. I examined a broad mix of authors and researchers who addressed a wide range of challenges, particularly concerning the embedment of sustainability and sustainable development. Recently reviewed publications included those with critical views on sustainability as well as those with perspectives from psychologists discussing people's behavioural patterns in the face of the climate crisis. Finally, I was inspired by professionals and academics known to me personally such as visiting lecturers to the ICIS Centre and Schumacher College, as well as British and Danish colleagues. The chapter on the three case stories derives from the case study from the PhD thesis, which was part of the investigation into sustainability and its practical application. Data was collected from the projects through interviews and constituted the field that I analysed in conjunction with data relating to the circumstances of each case.

In addition to the research described above the book is also an expression of my personal experiences working with sustainability for more than 30 years. My professional roles during that period were mainly as a leader. I also lectured and taught at a variety of venues on different continents, always in a sustainability context related to education, design, or organisational change: In addition, I developed and led several large EU-funded research projects within a sustainability framework. Parallel to my professional work I was interested in personal and spiritual development, as described in the Preface, which has added another dimension to the experience gained and my views. I believe that we need to evolve consciously and apply wisdom to what we do next to repair what we have broken and live in balance with ourselves, our fellow human beings and with all living species. We have shifted from being the observers to becoming the participants, and now we must cocreate the future—consciously.

1.4 Structure and Content

The book consists of eight chapters. The chapters can stand alone but together, they create a palette of insights that add up to the status quo and offer a kaleidoscope of factors for a trajectory of the future summarised in the final chapter. At the end of each chapter, you will find further reading and reference lists where I have included the reference material and other sources that you might find valuable.

Chapter 2

Chapter 2 deals with the essence of sustainability and provides an outline of the wide and multi-faceted field of sustainability, its origin, definition, interpretations, and key global implementation initiatives. It's essential to explore the area to provide a common foundation for an understanding of the construct and to allow for a deeper level of inquiry. This offers an opportunity to pinpoint critical themes and issues, which make the subsequent discussions of development models, utopian thinking, case stories and barriers, in the following chapters easier to digest as I return to these key themes throughout the book. There has not been a shared understanding of what sustainability is or means apart from the fact that currently, most people believe it is about environmental problems in general and climate change in particular. This view only covers a fraction of it's reach, as sustainability is much more than environmental problems and changing weather patterns. Education in sustainability has been lacking and is still not an inherent ingredient in the educational framework. This lack has left the field open to different interpretations and misinterpretations, many of which have added to people's resistence to implement sustainability. The second chapter, therefore, aims to address the original definition of both sustainability and sustainable development, set the context for their emergence and chart the first official initiatives introduced by the UN to further the agenda globally. This creates a common understanding of the book's main topics, and sets the basic parameters for the subsequent chapters.

Chapter 3

This chapter outlines the background and context out from which sustainability grew and it discusses the concept of development within the current economic framework. It deals with the pros and cons related to the present development trajectory that has advanced many nations, while the dark sides of the same coin have grown bigger to the detriment of other nations. In this process the natural environment, the biosphere and biodiversity have suffered to a degree that seriously undermines the foundation for the living species. In addition, human-generated global warming caused by the continuous amount of CO_2 being released into the atmosphere has created a situation where the impacts of the changing climate, are escalating out of control.

The chapter compares the current development model with the sustainable development model. Even though the word *development* figures in both models they are radically different. The chapter defines these differences and points to symptoms that indicate that there could be a shift away from the neoliberal capitalist paradigm brought about by the Covid-19 pandemic. The chapter discusses the state's role as the entity that is solely responsible for the well-being, safety, and progress of its citizens. As such, it is the most critical factor in implementing sustainability. Finally, the chapter looks at future perspectives and assesses the possibility of sustainable development within a business-as-usual framework.

Chapter 4

Chapter 4 examines alternatives to the current growth model and delves into utopian thought and ideas. It offers a look into history, from Plato's the *just* man and the *just* society to the present time's small intentional community and ecovillage ideals. The chapter examines the impact of both utopian as well as dystopian thinking on societies. Utopia is relevant because sustainable development is a profoundly different societal model to the one currently in place. Utopian ideas are radical and while conventional theories of development may have similar aims to improve the lives of citizens, it is from utopian theories we can learn about the nature of change. The experiments and how they have succeeded or failed in impacting societies over time offer us some practical insights. Linking this to the sustainability construct and the attempts at translating it into practical application utopia can also provide contexts and knowledge to learn from, including the barriers that are preventing change from happening.

Chapter 5

Chapter 5 discusses implementing sustainability in the current paradigm. The chapter documents key implementation initiatives and assesses the overall achievements and learning harvested from both top-down and bottom-up global undertakings since 1987. A great deal of research has been focused on the development of a variety of application methodologies in the form of principles, models, methods, indicators, monitors, and assessment tools. As the world woke up and started to understand the seriousness of the problems, the sustainability discourse became an item on most corporate, business, and political agendas. Scientific as well as other academic research into new technologies started to flourish. Many organisations, research institutions, NGOs, and pioneering individuals worldwide worked with the concept and developed frameworks and tools that could be used for sustainable development activities. The chapter looks at these endeavours and discusses some of the key implementation methods that grew, mainly from academic research and the business sector.

Numerous models have seen the light of day since the Rio Summit 1992, and new models are continually being added to the list. Among the most critical documented in the chapter are Corporate Social Responsibility (CSR), The Triple Bottom Line (TBL), The Natural Step (TNS), The Ecological Footprint, Cradle to Cradle (C2C), Circular Economy, and Permaculture. Recently the UN 17 Sustainable Development Goals have been added to the list. The chapter offers examples of assessments and standards to give an idea of the wide spectrum of implementation attempts and the complexity that they contain. The examples provided in the chapter indicate the multi-layered approaches they represent. Feedback from the application attempts is documented and offers an insight into the type of issues surrounding the initiatives as well as the positive and negative impacts the models have had on sustainable development.

Chapter 6

This chapter contains three case stories: the Danish island of Samsø, Brahma Kumaris (an international faith community for peace), and Novozymes (a Danish multinational company specialising in the production of enzymes). The descriptions of the cases are based on data from the interviews I held with the project development teams. The main focus is on the specific process of each case and covers the successes as well as the barriers or problems encountered. The three cases are very different. The diversity comprises culture, identity, vision, context, size, and the reasons for engaging in the initiatives for change. The chapter demonstrates sustainable development in action in a local and specific situation. It tells us what can be achieved in the current framework, regardless of whether the undertaking is an island, a faith community, a corporation, or a similar entity. It is due to initiatives such as these that the sustainability agenda has made some progress since 1987. The chapter's positive perspective is the knowledge that necessary changes can be made despite barriers, adversity and difficulties, provided that the motivation and determination are present.

Chapter 7

Chapter 7 defines key barriers hindering the transition to sustainability and expanding the discussion of the realisability of sustainable development. One part discusses the overall or universal barriers that are innate in our current societies, covering political, economic, technological as well as cultural aspects. The other part discusses the barriers uncovered in the case stories. Some barriers or obstacles are clearly defined and visible; others are subtle, indirect, and complex. An observation from my investigation was that the closer I examined the sustainability construct in the context of the existing societal framework, the more barriers I uncovered to its realisation. The barriers appear to be as interlinked and interdependent as the sustainability factors themselves. Some are broad, challenging obstacles, and some are practical and basic, relatively easy to understand, and address. The difficulty closing the gap between the theory of sustainability and its practical application does, therefore, not seem to be due to a lack of scientific research or technological inventions and know-how but due to the magnitude and amount of barriers found everywhere in our society. Some resistance is universal and encompasses the main aspects of our cultural and traditional make-up. Some is specific and locally orientated, such as the resistance experienced and recorded in the case stories. The chapter discusses each group of barriers as well as ideas to overcome them.

Chapter 8

The last chapter discusses future perspectives from a belief and viewpoint that a transition turning has started. The chapter highlights the challenging process of creating a state of global sustainability. It outlines potential frameworks and factors needed for it to be achieved and draws a trajectory for the future based on the sum of the diverse parts. In doing so, the chapter summarizes the key ingredients of the challenge based on the previous chapters and concludes with final reflections on its realisation. Questions relating to undertaking initiatives for change and overcoming barriers are addressed from the perspective of the current state of sustainable development. Utopian theories, historical examples, and practical experiences from the cases are summarized. New influences are added, including recent global events and calls for action, the impact of the COVID-19 pandemic, the IPCC 6th Assessment Report from 2021/2022 as well as agreements made at COP 26 (Conference of the Parties). The aspect and importance of *inner sustainability* comprising the emotional, psychological, and spiritual dimensions are discussed as they play a crucial role in the change of direction and in building resilience to meet future challenges.

Readings and References

Blincoe KS (2020) Essential sustainability: from utopian theory to practical application. Dissertation, Kingston University London. https://eprints.kingston.ac.uk/id/eprint/51146. Accessed 1 March 2021–1 March 2022

Broadbridge E et al (2021) The school for life: N.F.S. Grundtvig on the education for the people. Aarhus University Press, Aarhus, DK

Chora Connection (2019) Chora 2030, Copenhagen. https://vimeo.com/choraconnection. Accessed 10 Dec 2019

Gibran K (1958) The prophet: the voice of the master. Richard Clay Ltd., Suffolk

Goodwin B (ed) (2001) The philosophy of utopia. Routledge, Abingdon

Goodwin B (2007) Nature's due: healing our fragmented culture. Floris Books, Edinburgh, p 177

Goodwin B, Taylor K (2009) The politics of utopia. International Academic Publishers, Bern

Hauschild MZ et al (eds) (2018) Life cycle assessment: theory and practice, 1st edn. Springer Nature, Switzerland

IPCC Reports (2020) IPCC. https://IPCC.ch/reports/. Accessed 20 April 2020

Our Common Future (1987) UN. https://sustainabledevelopment.un.org/content/documents/5987our-common-future.pdf. Accessed 10 Jan 2021

Plato, Butler-Bowden T (eds) (2012) The republic: the influential classic. Oxford University Press, Oxford

Schumacher College (2022) Dartington. https://campus.dartington.org

Science for Environment Policy (2021) European Commission. https://ec.europa.eu. Accessed 20 Oct 2021

Stockholm Resilience Centre (2021). https://www.stockholmresilience.org. Accessed 5 Nov 2021

Sustainable Development Goals (2015) UN. http://www.un.org/sustainabledevelopment/sustainable-development-goals/. Accessed Dec 2015

UN (2021) United Nations. https://www.un.org/. Accessed 4 Feb 2021

6th Assessment Report (2022) IPCC. https://www.ipcc.ch/assessment-report/ar6/. Accessed 28 Feb, 4 April 2022

Chapter 2
The Essence of Sustainability

Abstract This chapter outlines the wide and multi-faceted field of sustainability, its origin, definition, interpretations, and key global implementation initiatives. Defining the field is essential to provide a common understanding of the construct and allow for a deeper level of inquiry into the essence of sustainability. This offers an opportunity to pinpoint key themes and issues which make the subsequent discussions of development models, utopian thinking, case stories, and barriers in the following chapters more relevant as I return to these critical themes throughout the book. There is no shared understanding of what sustainability is or means, and many people believe it deals with environmental problems in general and with climate change in particular. However, this only covers a fraction of the broad spectrum of its reach. Education in sustainability has been lacking and is still not an inherent ingredient in the educational framework. A lack, that has left the field open to varied interpretations many of which have added to the resistance to implement the principles and aims of sustainability. The objective of the second chapter is, therefore, to address the original definition of both sustainability and sustainable development, define the context of their emergence, and chart the first official initiatives that were taken to further the sustainability agenda globally.

2.1 Context and Emergence of Sustainability

> Sustainability is an all-encompassing construct. It is multi-dimensional, multi-disciplinary, transnational, transcultural and essentially holistic. It is the ultimate utopian model for a balanced world and one that can sustain human life for generations to come without destroying the biosphere and ecosystems.

Sustainability is essentially a new vision for the future of humanity. A vision we sorely need to be realised. The sustainability construct was first defined in the Brundtland Report in 1987 (Sect. 2.2) as an agenda to be pursued worldwide and is a relatively recent neologism and concept. However, some of the individual parts that make up the concept, for example, the environmental, social, and ethical dimensions, are not new in historical contexts. Most societies have in the past been concerned with social challenges relating to human development and well-being. Environmental

K. Blincoe, *Achieving Sustainability: The Ultimate Human Challenge*,
https://doi.org/10.1007/978-3-031-10023-9_2

challenges such as deforestation, soil erosion, drought, crop failures, polluted rivers, and ethical challenges such as justice and rights related to humanity and nature have existed for centuries. The author, Jared Diamond, describes in his book "Collapse: How societies choose to fail or survive" (2011), how advanced societies in the past have declined and disappeared due to environmental and resource problems. These societies outgrew their ability to sustain continued growth through their need for more resources, thereby generating scarcity and subsequent collapse. We are faced with a similar situation on a global scale. To address the problems, the constructs of sustainability and sustainable development were created and launched by United Nations in 1987.

Sustainability. What does it really mean? How is it interpreted? How has it been approached? Many attempts to adapt the construct to fit within diverse contextual settings in mainstream society have been made since 1987. However, the attempts to apply it have not produced the needed results. The aim of this chapter is to explore the essence, the wide reach, and the many facets of the concept and to assess its achievability. In the design world where I have been operating for some part of my career, it is acknowledged that the solution to a problem is often found within the problem itself (Pentagram 1978). The design process can simplify complexities and create solutions to many kinds of issues. If a problem is examined thoroughly from a variety of perspectives, the task to solve it becomes clearer and the right solutions emerge which can be developed to encompass relevant aspects and essential points. It is with this mindset the challenge to achieve sustainability is approached in this book.

Sustainability and sustainable development were developed by the UN as a global approach to combat the adverse environmental and social effects of industrial development and globalisation—the shadow or the dark sides of economic progress. The broad issues to be addressed represented a complex challenge given the current growth paradigm into which sustainability was introduced. This paradigm is in this book defined as an economic growth paradigm based on neoliberal principles and values that were introduced during the reign of UK Prime Minister Margaret Thatcher and the American President, Ronald Reagan in the 1980s. The neoliberal project was both economic and political and focused on principles such as free market policies, including international trade, free flow of capital, deregulation of markets, reduction in taxes and public spending. In addition it advocated an increase in privatisation, and a flexible labour market. It was basically a belief in supply and demand where the market forces were allowed to decide the policies of the country, and where government played a secondary role. Most countries worldwide adopted these policies, and although new political ideas to change direction started to emerge after the last financial crisis in 2007–2009, the neoliberal ideas have largely remained intact. I deal with this topic in more detail in Chap. 3, but it is out of this context that sustainability emerged, and sustainable development was introduced.

When discussing context, it is necessary to address the role of the state. The state is the principal change agent, implementor and guarantor for securing development that benefits all citizens of a nation. The state is, in this connection, understood to be the entity with the legitimacy and power to influence and control activities such as public service, distribution of income, the flow of money, investment, and the work force. The state can also enter into agreements with other states if global partnerships are needed. The state's role is crucial in securing the well-being of a nation, as we have experienced during the Covid-19 crisis. The impact of the state promoting the economic development trajectory, mainly due to the belief that it had a positive and qualitative impact on humanity, has been profound, with both positive and negative outcomes (Kingsbury et al. 2016).

Adverse Impacts of the Growth Paradigm

Significant adverse impacts of the growth paradigm leading to environmental degradation, greater inequality, and threats to the planetary boundaries began to be noticed in the eighties. Symptoms of global warming and climate change had not until then been fully understood, communicated, and experienced in the mainstream and were not regarded as sufficiently grounded in science. However, the UN IPCC assessment reports, as well as special climate change reports, published by the UN from 1990 onwards, became more widely publicised. Increasing scientific evidence pointed to changes in weather patterns and the fact that it was human-induced. The awareness spread and the data gained attention. At the same time, more widespread impacts of the changing climate started to be felt around the world through extreme storms, flooding, droughts, deforestation, and forest fires, and the time issue began to be an essential factor.

The twentieth and twenty-first centuries have seen researchers, writers, and pioneers in all areas of society, from academia, business and civil society to politics, examine the escalating negative consequences of economic and industrial development. It is no wonder that the present epoch or geological time has been called the Anthropocene as the (unofficial) name given to the recent period of the history of the earth, signifying the fundamental impact of human activities on the planet. Significant early twentieth-century authors such as Aldous Huxley with the book "Brave New World," first published in 1932, Rachel Carson with "Silent Spring" from 1962; and James Lovelock with his series of books about Gaia (the living planct), which he published between 1995 and 2019 and many others, gave critical views of the consequences of the political and economic policies at that time. They alerted people that the consequences of development posed a danger to humanity if not addressed. The mounting evidence of the emerging problems led the UN to act, starting with the Stockholm Convention in 1972 on environmental issues. It was the first world conference where the environment was the key topic for discussions. Authors and visionaries continued to support the discourse during the following decades. One significant input was made by former Vice President Al Gore. He described and demonstrated graphically and powerfully in his book and lectures, as well as in the film "An Inconvenient Truth" from 2006 how the exponential increase in the use of non-renewable resources and energy matched the equally exponential consumption and CO_2 emissions on a planetary scale. This lecture was a worldwide communication breakthrough. The brilliant visual imagery of the work affected many of us, and the dramatic increase in global warming became clear and visible to all. This element has been sorely lacking in the environmental discourse as most scientific presentations, diagrams and visual imagery are complex, difficult to decipher and understand.

A few years later, at a conference briefing in Copenhagen, in the spring of 2010, both Connie Hedegaard, the then EU Commissioner on climate and Jacqueline McGlade, the then Director of the EEA (European Environment Agency), warned of climate change occurring at our doorstep. Jacqueline McGlade considered Europe a metaphor or microcosm of the global changes. They both predicted a Europe with areas of extreme drought desertification and severe flooding with significant differences between North and South. Considering the drought of 2017, the flooding in Central Europe in 2021 and the recent heatwave with temperatures reaching into the 40's in 2022, they were right. Climate change is impacting us all with unfortunately the poorest countries suffering the most.

A Radical and Necessary Change

Sustainability and sustainable development were introduced as a remedy to adjust and repair the imbalances. But closing the gap between the ideal world and the real world has proven difficult. It is debatable whether sustainability, as defined in the Brundtland Report (see the section below), is implementable given the strength of reigning economic paradigm. Or should it instead be regarded as a long-term theoretical and utopian ideal for a better society? Utopian visions for the perfect nation, community, or world have always existed. Throughout history, humans have had the ability to envision and describe their ideal environment. Sometimes they have remained dreams and ideals, and sometimes they have been tested in real-time. The ideals have not changed significantly over time. The vision of a state of sustainability is comparable to the one described in Plato's "The Republic" from around 380 BC. "The Republic" promoted a different and radical worldview from the existing beliefs. It addressed societal problems of that time and advocated new ideas for systems changes, in particular the order and character of the *just* state and the *just* man. Sustainability promotes similarly a different and radical worldview with an ethical core that requires radical change. Both constructs are rooted in social, ethical, and moral principles. Plato's "Republic" is regarded as one of the most influential works of philosophy and political theory. Its ideas and ideals are still discussed today in various social and political contexts. The question remains whether or how the gap between a ground-breaking ideal and a practical reality can be bridged. What kind of intervention is needed? This link between utopian constructs and sustainability is discussed further in Chap. 4.

In its essence, the concept of sustainability is ground-breaking. It is intended to be both holistic and inclusive. John Blewitt, who is an expert on sustainable development from the Aston Business School in the UK, writes in his introduction to one of his books on "Understanding Sustainable Development" (2008):

> Sustainable development requires an understanding of the natural world and the human as being not so much 'connected' as one and the same.

And he writes further:

> It [sustainability] requires a holistic way of looking at the world and human life.

This perspective on sustainable development is one that I share, not only due to the theoretical research for the PhD on which this book is based as discussed in the introduction but mostly from working with sustainability in a practical context for nearly 30 years, directly experiencing the challenges of bringing the construct into real life.

The current paradigm that dominates our world is based on an atomistic worldview that promotes fragmentation (Goodwin 2007) and does not regard the world and human life as *one and the same.* It encompasses the view that material progress is the ultimate goal, and that man is separate from and greater than nature. Holism is rarely seen as part of the discourse, especially at high levels of influence. Working with an ideal or concept where humanity and nature are regarded as equal and not only connected but believed to be the same does imply an integration that has no room for separation or fragmentation. The gap between theory of sustainability and practice of sustainability is therefore experienced as being difficult if not impossible to bridge. However, to further understand the deeper issues related to sustainability and sustainable development it is important to revisit the original meaning of the concepts.

2.2 Brundtland Definition

The word sustainability originates from the Latin sustinere. Sus = up, tenere = to hold. Sustainability is in the dictionaries explained or defined as the ability to *sustain*, to *endure*, to *maintain.* The word has many definitions, depending on the context in which it is seen. But all of them, fundamentally, refer to what is accepted as the classic definition, the Brundtland definition from 1987.

The first official use of sustainability to describe the context and link between environmental, economic, and social issues as we understand it today was the Brundtland Report: "Our Common Futurc", publishcd in 1987. The Brundtland Report was the result of the work of the Brundtland Commission, originally called the World Commission on Environment and Development (WCED), set up by the UN General Assembly in 1983. The commission was set up to address the growing concerns about an increasing deterioration of the environment and natural resources and the related negative impacts on economic and social development. Gro Harlem Brundtland, the former Norwegian Prime minister who gave the name to the report said the following in the introduction of the report:

> Environmental degradation, first seen as mainly a problem of the rich nations and a side effect of industrial wealth, has become a survival issue for developing nations. It is part of the downward spiral of linked ecological and economic decline in which many of the poorest nations are trapped. Despite official hope expressed on all sides, no trends identifiable today no programmes or policies, offer any real hope of narrowing the growing gap between rich and poor nations. And as part of our "development", we have amassed weapons arsenals capable of diverting the paths that evolution has followed for millions of years and of creating a planet our ancestors would not recognize (Brundtland 1987).

Some of these words still sound familiar more than three decades years later. However, were Gro Harlem Brundtland to write this today I am sure the words would be graver, more alarming, and more urgent as we have continued to undermine and aggravate the conditions for our continued survival rather than course correct as the Brundtland report recommended. The commission was asked to formulate "A global agenda for change". It established that environmental problems were essentially global and suggested it was in the common interest of all nations to develop policies for sustainable development. The Brundtland Commission formed and defined the concept of sustainable development to mean that economic, as well as social development is necessary to secure a healthy and productive life for people. But it had to be a development that did not compromise the ability of future generations to have healthy and productive lives. The Brundtland concept of sustainable development thus addressed and included four important elements:

- Human-related issues including basic needs (food, water, housing, education, and health).
- Social and ethical issues relating to life quality and dignity.
- The concept of limitation. There is a limit to non-renewable resources, and the present generation must limit the use of these to ensure that there are sufficient resources for future generations.
- The concept of natural resources being vital to sustain life.
  (WCED 1987)

The two pillars that underpin the sustainability definition are needs and limits. There are needs to be met globally, especially in the poorest countries where people lack basic necessities such as food, clean water, and sanitation, and there are limits to the natural resources available for human consumption, most of which are being used beyond the level of necessity. The essence of sustainability is therefore essentially an ethical issue: to secure the well-being of all people on the planet by meeting their basic needs and doing this within the resource limits available, making sure that these are used responsibly and that the natural habitat remains healthy.

For the first time, it was publicly announced that men and women should live in balance with nature, so that future generations could have the same opportunities and possibilities as people have today. The core of sustainability is balance. A balance between the material and other demands that humans make on the earth and the earth's capacity to sustain such demands both in the short as well as in the long-term. To achieve this balance, all the above factors must be taken into consideration. The underlying unique value contained in the concept is the ethical element. This aspect becomes apparent in the context of environmental, social, economic as well as political issues due to the moral values that underpin every activity related to sustainability (Kibert et al. 2012). A parallel to seeking to achieve the *just* state or the *just* society that Plato advocated where the good of the individual was equated with the good of society.

From a rational perspective, we need to rethink how we live as we have a finite planet, and the population continues to increase. Capitalism clearly does not support sustainable development as the capitalistic goal is continued economic and material growth, mainly benefitting the rich. However, the Brundtland definition does have within it some clues as to why implementation of sustainability has proven to be challenging and not only from an economic point of view because it is holistic and ethical in its essence and advocates a framework for a different paradigm to the one we have today. Thus, change is required, and not only a change to the economic system but a deep transformational change of our societies that require both systemic, structural, and operational interventions. We fundamentally need new constitutions.

In this context elements parallel to utopian ideals and visions of an alternative future can be explored and be informative. Some indication of the probability of achieving the goal of sustainability can also be found looking back to similar turning points and paradigm shifts in history. There are many examples of failures to change direction, as mentioned previously, where I referred to Jared Diamond's "Collapse". If Diamond is right, humanity is currently on a decline, growth will cease, and societies will disappear. James Lovelock mentioned earlier, supports this in his book "Novacene" (2019) where he predicts that only a small percentage of people may be left to inhabit the earth by the end of the twenty-first century. This scenario is also addressed by the writer and journalist Wallace-Wells, who published an apocalyptic book in 2019 called "The Uninhabitable Earth: A Story of the Future", where he said: "In fact, we are only just entering our brave new world, one that collapses below us as soon as we set foot on it".

I will not, in this book, argue for the likely survival of humanity, nor will I argue for the opposite. One part of me thinks that the human species has evolved its need for materialism too fast to become resilient and that we probably will become extinct within the next century or so, depending on the degree to which we continue to erode the base that sustains us. Another part of me stands back and looks at all that humanity has created, and I believe that we ought to be capable of making our lives sustainable, regenerating, and re-wilding nature and learning to live in harmony with it. If we fail, it is not that we cannot; it will be that we will not. We have the knowledge, the technology, the capacity, and the financial capability necessary. Resilience not only means robustness in terms of survival; it also means developing or accessing our wisdom and expanding our consciousness. Creating a sustainable life is the obvious next evolutionary step for humanity. Human psychology plays a major part in applying the will to do what must be done. If there is no will and there is a persistent hanging on to the status quo and business as usual, nothing will change. When the Covid-19 pandemic hit, most governments chose to put peoples' lives first rather than pursuing economic growth. It was an example of a clear choice and focus even though the financial repercussions could have been more far-reaching than we experienced. Some countries suffered more than others, but overall, some level of global resilience seems to have been supported through government intervention.

Interpretations and Misinterpretations

Sustainability is often confused with other related concepts such as environmentalism or ecology. There has been a lack of clarity regarding the meaning of each. Many authors in books on ecology and sustainability have mixed the three concepts, making a complex field even more complicated for lay people to comprehend. Understanding is essential for problem-solving. Confusion leads to misconceptions, myths, and fake news. It is often difficult to know what people mean when they talk about sustainability especially when they confuse the terms. The definitions, for example, of the three concepts mentioned are easily accessible and are fairly clear. The dictionaries Collins (2012) and The Concise Oxford Dictionary (1982), as an example, define the concepts in the following way:

- *Ecology* is a scientific or biological study of the interrelationship of living organisms between themselves and with their environments. Ecological is understood to mean beneficial to or not harmful to the environment.
- *Environmentalism* is described as a theory that views the environment as the most important or primary factor in human development and is concerned with the protection or quality of nature. Environmentalism is also described as the process of protecting the ecological balance.
- *Sustainability* is the capacity to endure and sustain life. It is understood to maintain society at a balanced level without exhausting natural resources or damaging the environment.

The concepts have the environment as a commonality and as an essential component in their meaning. Where ecology focuses on the natural systems and their interrelationship, environmentalism deals with the protection of the natural habitat as it is seen to be critical to human development. Sustainability includes the other two concepts and encompasses both the interrelationship of the natural systems as well as the protection of it. It also includes additional dimensions apart from the environmental and is thereby more encompassing than environmentalism and ecology. In there also lies its uniqueness and complexity.

The above explains why phrases like *ecologically sustainable* and *environmentally sustainable* are used colloquially. It may be that the mixing of the terms can either be attributed to authors not knowing the different definitions or that their contexts and viewpoints have a leaning towards the environmental or ecological aspects of sustainability and should be seen from that perspective. This confusion and lack of clarity can be and is often used as an argument against sustainability. Frequently I hear people talk of finding another word to replace *sustainability*. Their argument is that the term is used to describe contexts that are not sustainable such as *a sustainable economy*.

Another similar example is the word *ecosystem,* which appears to have been adopted by the business sector to describe an industry. McKinsey's annual report (2021) on the fashion industry repeatedly uses the word *ecosystem* to talk about the success of the industry not in a sustainability context but purely as an economic survival mechanism. No wonder we get confused and irritated. The world of communication is dominated by social media consultants and an advertising industry who know what to do and say to continue business as usual.

There has also been confusion around the difference between the term *sustainable development* and *sustainability*. Some researchers such as John Blewitt and Helen

Kopnina state in their book "Sustainable Business" (2017) that the term *sustainable development* is an "oxymoron" as the term sustainability means stability and endurance and that development signifies "change" and "dynamism". From my perspective development is a dynamic movement towards something that has not yet been achieved but is still growing and unfolding—a continuous process. As such sustainable development can be interpreted to mean a movement or a development towards a state of sustainability. The two terms can simply signify two different states.

The intention of the UN promoting sustainable development was as a process to be undertaken by the world nations, led by the rich countries to move towards a global state of sustainability. Sustainable development can therefore be viewed as a trajectory towards a steady state of sustainability. The real problem seen from this perspective is that the general understanding of sustainable development is as a reformist development theory. A kind of business as usual with a few changes although still promoting economic growth at the same time as avoiding environmental degradation, pollution, etc. However, as discussed earlier this is a misconception as a state of sustainability calls for a paradigm shift. Continuing the economic growth trajectory is not compatible with reducing CO_2 emissions or increasing biodiversity—it is incompatible with sustainability.

2.3 Sustainability in a Development Context

> Humanity has the ability to make development sustainable to ensure that it meets the needs of the present without compromising the ability of future generations to meet their own needs (WCED 1987).

The concept of sustainable development is not new. It was first written about in 1980 by the World Conservation Strategy (WCS), which was set up by the IUCN (International Union for Conservation of Nature) to find practical and rational solutions to environmental and developmental challenges. IUCN is the world's oldest and largest environmental organisation, focusing on conservation at a global level.

The definition of sustainable development is: "Development that meets the needs of the present without compromising the ability of future generations to meet their own needs" (UN 1987). Sustainable development is therefore a process of developing the ability of humanity to live and work within the carrying capacity of the natural habitat. It is about balancing the social, economic, and environmental aspects of society, as discussed in the section above. As such sustainable development means almost the opposite of the conventional view of what development currently exemplifies and symbolises.

One of the three main interlinking factors of sustainable development is the environmental factor. Natural resources and the natural habitat are considered essential for sustaining life itself. The recent model to achieve sustainable development, the 17 Sustainable Development Goals (UN 2015), separates environmental initiatives from social and economic goals. It breaks up the interdependent 3-factor model into 17 separate goals. The reason for this fragmentation is due to wanting to ease and simplify the issues of implementation, making each of the seventeen goals more accommodating and accessible, more easily achievable, and less challenging to address. Inherent in this lies the key contradiction and challenge of the 17 SDGs, as they only make sense in a sustainability context if all goals are met. Regardless of whether the attempt is to balance the social, environmental, and economic factors of the Triple Bottom Line model (Sect. 5.2) for sustainable development or to meet each of the 17 Sustainable Development Goals, undertaking this task is challenging—and may not be possible to achieve with the current mindset.

2.4 Key Global Initiatives

Consideration the above, it is therefore interesting to examine some of the initiatives and attempts to promote and implement sustainable development after 1987. Then to look at why so little has been achieved. Below are some of the key events.

Earth Summit—Rio 1992

The birth of sustainability as a publicly acknowledged concept for a new development direction started with the United Nations Conference on Environment and Development (UNCED), also called the Rio de Janeiro Earth Summit, which took place in Brazil in June 1992. The world governments were invited to participate, and 172 countries signed The Rio Declaration on Environment and Development, which covered 40 principles dealing with a variety of environmental and social issues. Other outcomes from the summit included the Framework Convention on Climatic Change, which promoted research, regular meetings, and discussions relating to policy agreements on climate change, food production, economics, and related issues (UN 1992). In addition, Principles of Forest Management were drawn up, which included recommendations on conservation and forestry management and development, and the Convention on Biological Diversity, which focussed on biological diversity, resource usage, and fair sharing of benefits. One of the most well-known initiatives was Agenda 21.

Agenda 21 was a global non-binding plan for action where member countries could engage national and local government and civil society to start and participate in activities related to sustainable development. The initial ambition for the summit in 1992 was to achieve sustainable development globally by 2000. In the light of today's climate and sustainability discussions, the target of 2000 produces a feeling of unease. The fact that it was not achieved should alert us to question whether the current ambitious targets of CO_2 neutrality by 2050 are realistic.

The Kyoto Protocol

> The Kyoto Protocol was the first critical step – today we must take further and more far-reaching action toward a truly sustainable future for seven billion, rising to over nine billion people (UNFCCC 2015).

As an extension of the UN Framework Convention on Climate Change (UNFCCC) from the Rio Summit, the Kyoto Protocol was set up in Kyoto, Japan in 1997. It was an international treaty that committed countries to reduce greenhouse gas (GHG) emissions, acknowledging that global warming was occurring and that CO_2 emissions were caused by human activity. Countries that signed the agreement committed to reducing emissions by 5% of the 1990 levels by 2012. As the developed countries were recognized as responsible for the majority of emissions, it was thought that they should bear the heavier burden of the reduction. This view was also due to the fact that developing countries could not afford to switch to cleaner fuels or renewable energy. It was the intention, that with time the emerging countries would be included in the protocol, as cleaner technologies would be integrated and made available to them at lower costs by the rich part of the world (Caradonna 2016).

The framework for the implementation was agreed upon in 1997. The Kyoto Protocol was activated in February 2005, with the first commitment period set to start in 2008, and to end in 2012. Even though 36 countries from the developed continents managed to reduce emissions according to the protocol, the overall global emissions continued to rise—32% in the 10 years from 1990 to 2010. An extended commitment was then agreed on in Doha in 2012 with a deadline of 2020. Despite several conferences and meetings, little progress was made as many faulted the initiative. The most straightforward interpretion of the results after the protocol had been in effect for ten years was through the comments from Christiana Figueres, the past Executive Secretary of the UNFCCC. She stated: "Despite our best efforts, greenhouse gases continue to rise, threatening sustainable development and putting millions if not billions of people at risk over the coming decades" (UNFCCC 2015).

This sentiment is also indicative of the results of more than two decades of work on global initiatives. It adds weight to the argument that sustainability will take longer to become an integrated part of our societies than people expect and hope.

Earth Summit: Rio + 10

Ten years after the Earth Summit in Brazil, the Rio + 10 Summit took place in Johannesburg in South Africa. It was a follow-on from the summit in 1992, and one of the key outcomes was the Johannesburg Declaration on Sustainable Development. It was an expansion of the declarations from 1972 in Stockholm and 1992 in Rio emphasizing the threats to sustainable development and encouraging multilateralism and partnerships. The declaration was not legally binding and was more general, for which it was criticised by increasingly aware global NGOs and other environmental organisations. One lasting result from the summit was an acknowledgement of the interlinking of the three factors of sustainability: the environmental, economic, and social factors, that since then have become an integrated and accepted element of sustainable development (Farley and Smith 2014).

Earth Summit: Rio + 20

The Rio + 20 Summit in 2012 was a critical global event attended by around 180 world leaders, other negotiators, and approx. 50.000 delegates. The official aims of Rio + 20 included promoting "a green economy in the context of sustainable development and poverty eradication" as well as the "institutional framework for sustainable development" (UN 1992). One of the aims of the event was to assess the progress achieved following the first summit in 1992. Furthermore, to discuss the opportunities for advancement and examine how Rio + 20 could build on the opportunities and challenges to create a more ambitious action plan. The world leaders signed a 283-point declaration called "The Future We Want" which was aimed at steering the world towards sustainable development.

In assessing the progress made over the 20-year period between 1992 and 2012, both positive, as well as negative feedback was publicised regarding the outcome. The criticism included that the declaration 20 years on lacked measurable commitments, specific goals, and timelines for further progress. No legally binding commitment was made—a similar situation to the declaration in 1992. The positive feedback given by politicians and the business sector was that the declaration ensured a continuation of the process towards sustainable development—albeit on their terms. Governments and organisations speaking for the developing countries that suffered the most from climate impacts were disappointed and frustrated. The pattern that emerged from the congresses and summits is that throughout the period from the Rio Conference in 1992 until today governments have met, got inspired and signed agreements and declarations but have failed to fulfil their promises.

United Nations Sustainable Development Summit

Two significant events took place in 2015: United Nations Sustainable Development Summit in September in New York and COP 21 in December in Paris. The first gathering of world leaders was to agree on the 17 Sustainable Development Goals at the UN General Assembly. At the second event, the UN Climate Conference, the Paris Agreement was signed on reducing global warming.

The 17 Sustainable Development Goals

The 17 Sustainable Development Goals (SDG) included 169 sub goals and 239 indicators "to end poverty, protect the planet, and ensure prosperity for all as part of a new sustainable development agenda". The 17 SDGs were an extension and an expansion of the Millennium Goals (see below). Specific targets were set to be achieved over the following 15 years until 2030. One of the new and different aspects of creating the goals compared with previous initiatives was the fact that the SDG's were developed by a diverse group of actors, including representatives from governments, the business sector as well as civil society. The UN stated that to achieve the targets in the required timespan, the involvement of all sectors of society should work together. Thus Goal Nr. 17 advocating partnerships became one of the key focal points of the model.

The Development Goals were an extrapolation of the Millennium Development Goals. These were established at the Millennium Summit in 2000 and comprised eight socially orientated targets including anti-poverty targets to be accomplished by 2015. Progress according to recent statistics was made towards achieving the goals. For example, it was stated in the 2018 IPCC report and in the book on world statistics "Factfulness" from 2018 written by the Swedish physician and academic Hans Rosling that global poverty was on the decrease and had been for some time, especially in Asia. Nearly all children (95%), both boys and girls, were now attending primary school; the number of child deaths was decreasing; several diseases were being treated (AIDS, Malaria, and Tuberculosis) with positive results. One of the reasons for reduced poverty was said to have been the growth of countries like China, Brazil, and other developing countries. What was not known was to what extent increased global warming and climate change would offset this growth.

The 17 SDGs that followed on from the Millennium Goals were much broader, more wide-ranging, covered most aspects of society, and were regarded as the best model to achieve the needed transition. The question is: will the SDG model succeed in meeting the far-reaching goals that are required to turn the world from being unsustainable to sustainable? In the affirmative, there is a stronger impetus today in terms of scientific evidence of human-made global warming than there was in 1992. But on the downside, there is also increased evidence of the negative impacts of extreme weather affecting people everywhere as is the evidence of greater inequality.

Furthermore, there remains an estimated timeframe of approx. 10–20 years to reduce CO_2 emissions sufficiently to keep within the desired limits for global warming. Governments, business, and civil society are more aware of the perils of failing today than earlier. There seems to be a growing willingness and determination to change due to the increased stress on the planetary boundaries, and the general public is starting to show their frustration with politicians not taking action. This situation has not been so before. The scientific case for a trajectory toward a 3–5 °C increase in temperature is affecting decision-makers. Businesses are also starting to see an opportunity to shift from unsustainable to sustainable development. The election in Denmark, for example, in the spring of 2019 had climate change at the top of the political agenda and the wish to pursue a zero emissions strategy is stronger today than at any previous time.

The COP: Conference of the Parties

COPs (Conference of Parties) was set up by the UNFCCC in 1994 to address greenhouse gas emissions and protect the earth from climate change. The COPs are annual conventions where the world leaders are invited to discuss and make decisions regarding global warming. The first COP was held in Berlin in March 1995. The unique aspect of the COPs is that it is the only platform where the world leaders every year meet face to face to discuss joint issues. It is where north meets south; east meets west, and the developed countries stand face to face with the developing countries. They have an opportunity to discuss the issues close to heart, air their views and talk in larger as well as smaller forums. At the same event and outside the formal political discussions, more and more members of civil society from all around the world gather to discuss, demonstrate, and create diverse activities across the divides of gender, race, age, language, culture, and interests. The COP is a unique event. The United Nations is a unique constellation and organisation, being the only shared, global meeting place we have that includes all nations. Even if much criticism is directed at the organisation it plays an important role in mediating and allowing for communication rather than confrontation.

The COPs have the same purpose. It is here joint agreements are made across borders. Most agree that the results over the years haven't had the desired effect and that some members, especially the richest and largest countries have reneged on proper and binding commitments to curb global warming. However, the COP is the forum where delegates can jointly put pressure on those best placed to effect change, and where governments are forced to show their hand and meet face to face. There have been COPs where significant decisions have been made. However, not all good. Some of the more recent events include COP15 in Copenhagen in 2009, which produced the Copenhagen Accord. It was thought of as creating one of the most disappointing results of any of the conferences, as expectations had been very high leading up to the summit. Then there was the successful COP 21 in 2015, which produced the Paris Agreement on the reduction of CO_2 emissions to keep temperatures below 1.5 °C. The most recent COP 26 in Glasgow 2021 was the conference with the highest degree of awareness and commitment due to the intensifying climate crisis. Both COP 21 and COP 26 are described in more detail in the text below.

IPCC (The Intergovernmental Panel on Climate Change)

Leading up to the COPs, a massive amount of published scientific research documents is assessed and collated into IPCC reports. The IPCC was established in 1988 by UNEP (United Nations Environment Programme) and the WMO (World Meteorological Organization). IPCC publishes knowledge on human-induced climate change in the form of assessment reports, special reports, and methodology reports. The research and data that are collected are carried out by a large, diverse group of scientists from around the world and are produced to inform and advise the world's governments on climate change, policy, and specific issues. The work is scientific transparent, and is vital information for governments from which to develop policies. The reports provide the most relevant and up-to-date information and facts on climate change, providing strategies for adaptation and mitigation to reduce the risk of the impacts and to build resilience in our societies (IPCC 2022).

COP 21

The Conference of the Parties in Paris in November 2015 was the 21st event where the world's nations met to discuss climate change and related issues. It was a significant global event. The outcome was the Paris Agreement, a legally binding international treaty that aims for a joint global intervention to reduce climate change encompassing climate adaptation, mitigation, and financing. It replaced the Kyoto Protocol. One hundred ninety-six countries agreed to the text in the agreement. It was to be enacted when joined by at least 55 nations which had to represent at least 55% of global

CO_2 emissions. This was achieved in November 2016, and subsequently, countries began to develop plans and strategies for the reduction of emissions. Today, in the middle of 2022, a total of 193 countries have ratified the Paris Agreement. The key goal is to reduce CO_2 emissions to keep global warming 'well' below 2 °C with an intent to limit the temperature increase further to 1.5 °C compared with pre-industrial levels. However, with the latest findings, this no longer seems realistic. The long-term aim is to reach zero emissions in the second half of the century. Most nations have committed to NDCs (Nationally Determined Contributions) and to report on their results yearly.

COP 26

COP 26 was an equally critial venue to follow up on global commitments from the Paris Agreement. An additional aspect to this COP made it different from the previous conventions. The Covid crisis had an impact both in terms of the global future to be discussed and how the convention was organised. From a practical point of view, accessing the convention and the various events was difficult as tests and proof of vaccination had to be presented before attending, thereby placing a difficult administrative burden on the organisers and adding to the frustrations of the attendees both inside and outside the venues.

Most countries pledged to reduce emissions by 60–70% by 2030 and reach zero emissions by 2050. China pledged to cut CO_2 emissions by 2060 and India by 2070. The biggest disappointment was the failure to keep the wording of a commitment to stop coal production in the final agreement of the Glasgow Climate Pact. Instead, the wording *phasing out* was inserted, which upset many governments keen to stop the use of fossil fuels.

At COP 26, numerous agreements were made, and many partnerships and alliances were set up. Goals were clarified. Intentions flourished. Globally we are more informed about climate change and its effect than ever before. We are all impacted. More and more actors worldwide and from all walks of life are joining the fight for mitigating and adapting to the changing weather patterns, such as businesses, NGOs, financial institutions, public and private institutions, as well as organisations of all kinds representing minorities, nature, cities, culture, religions, youth. Radical policies and massive financial support are required. For the first time, these topics were discussed in depth at COP 26, and promises were made. More on this in Sect. 8.5.

We are faced with a situation, an enormous challenge that needs to be met within a short timeframe. The problems the world faces are concurrent to those in Diamond's examples, multiplied with the added factor of the ramification of the changing weather patterns. The situation is extreme and could bring about a collapse of the current way of living according to several researchers and writers. The state of the planetary boundaries (Sect. 1.1) points to a possible collapse of the natural systems within the foreseeable future unless the situation is redressed. We have known for several decades that we are moving towards a potential global catastrophe. The EU-2021 Strategic Foresight Report states that climate change is creating difficult living conditions for millions of people worldwide. It is estimated that there are currently 23 million people migrating due to climate change. The same report predicts that there will be approx. two hundred million fleeing extreme weather by 2050.

Sustainability is the only rational remedy to address the crises. From launching the concept in 1987 until now, some progress has been made. Discussions have changed from how to develop new principles, models, and methods to how to implement them and what policy decisions should be made by governments. Awareness of the issues has spread globally and locally in tandem with the impacts people have felt everywhere. However, the recent sobering statement by the UN Climate Secretariat tells us that "the sum of climate promises from the world's countries will increase the global CO_2 emissions by 16% by 2030. According to the Paris Agreement the emissions should be reduced by 45%" (UN Climate Secretariat 2021). We are not doing enough. The question is what is enough, and when and how do we know that what we are doing will achieve what is needed? The following chapters address different aspects of this challenge.

Readings and References

About Rio (Rio Declaration) (1992) UN. https://www.un.org/en/development/desa/population/migration/generalassembly/docs/globalcompact/A_CONF.151_26_Vol.I_Declaration.pdf. Accessed 5 Feb 2019

About the International Centre for Creativity, Innovation and Sustainability (2019) ICIS. https://sites.google.com/site/iciscenter/about-icis/history. Accessed 10 Jan 2019

Agyeman J, Bullard RD, Evans B (eds) (2003) Just sustainabilities: development in an unequal world. MIT Press, Cambridge US

Anderson RC (1998) Midcourse correction. Pereginzilla Press, Atlanta

Annual Report (2021) Mckinsey & Co. https://www.mckinsey.com/featured-insights/future-of-work/. Accessed 17 Aug 2021

Blewitt J (2008) Understanding sustainable development. Earthscan, London, p 1

Blincoe KS (2020) Essential sustainability: from utopian theory to practical application. Dissertation, University of Kingston. https://eprints.kingston.ac.uk/id/eprint/51146. Accessed 2020–2022

Brown LR, Gardner G, Halweil B (1999) Beyond Malthus. The Worldwatch Institute, New York

Brundtland Report, Our Common Future (1987) UN. https://sustainabledevelopment.un.org/content/documents/5987our-common-future.pdf. Accessed 10 Jan 2021

Business and the 17 SDGs (2019) UN Global Compact. https://www.unglobalcompact.org/. Accessed 5 Aug 2019

Caradonna JL (2016) Sustainability: a history. Oxford University Press, New York

Climate Change: Impacts, Adaptation and Vulnerability (2007) AR4 IPCC, UN. https://www.ipcc.ch/report/ar4/wg2/. Accessed 10 May 2019

Collins Dictionary (2012) Harper Collins, Glasgow

Concise Oxford Dictionary (1982) Oxford University Press, Oxford

Crutzen PJ, Stoermer EF (2000) The "Anthropocene." Glob Change Newslett 41:17

Daly HE (1996) Beyond growth: the economics of sustainable development. Beacon Press, Boston

Diamond J (2011) Collapse: how societies choose to fail or succeed, 2nd edn. Viking Penguin, New York

Diamond J (2012) The world until yesterday. Allen Lane, London

Doppelt B (2003) Overcoming the seven sustainability blunders. Syst Thinker ® 2(14):5

Eisenstein C (2013) The more beautiful world our hearts know is possible. North Atlantic Books, Berkeley

Ekins P (1992) A new world order. Routledge, London
Farley HM, Smith ZA (2014) Sustainability: if it's everything, is it nothing? Routledge, Abingdon
Figueres C, Rivett-Carnac T (2020) The future we choose. Manilla Press, London
Ghosh A (2016) The great derangement: climate change and the unthinkable. University of Chicago Press, Chicago
Gore A (2007) An inconvenient truth: the planetary emergency of global warming and what we can do about it. Bloomsbury, London
Goodall C (2007) How to live a low-carbon life. Earthscan, London
Goodwin B (2007) Nature's due: healing our fragmented culture. Floris Books, Edinburgh
Harvey M (2019) Utopia in the Anthropocene: a change plan for a sustainable and equitable world. Routledge, Abingdon
Hopper P (2012) Understanding development. Polity Press, Cambridge
IPCC Reports (2020). https://IPCC.ch/reports/. Accessed 20 April 2020
Jacobs M, Laybourn-Langton L (2018) Forum: a new economic paradigm. Paradigm Shifts Econ Theory Policy 53(3):113–118. https://www.intereconomics.eu/contents/year/2018/number/3/article/paradigm-shifts-in-economic-theory-and-policy.html. Accessed 7 May 2020
Kibert J et al (2012) Working toward sustainability: ethical decision making in a technological world. Wiley, Hoboken
Kingsbury D et al (2016) International development: issues and challenges, 3rd edn. Palgrave, London
Kirby J, O'Keefe P, Timberlake L (eds) (1995) The earthscan reader in sustainable development. Earthscan Publications Ltd., London
Klein N (2014) This changes everything. Simon & Schuster, US
Limits to Growth (2019) Club of Rome. https://www.clubofrome.org/news/publication-of-the-limits-to-growth/. Accessed 10 Oct 2019
Lovelock (2019) The Novacene: The Coming Age of Hyperintelligence. Allen Lane, UK
Meadows DH, Randers J, Meadows D (2004) Limits to growth: the 30-year update. Chelsea Green Publishing Company, White River Junction
Millennium Goals (2000) UN. https://www.who.int/topics/millennium_development_goals/about/en/. Accessed 6 Aug 2018
Orr D (1994) Earth in mind. Island Press, Washington
Papanek V (1995) The green imperative. C.S. Graphics, Singapore
Pentagram (1978) Living by design. Lund Humphries Publishers Ltd., London
Planetary Boundaries (2015) Copenhagen University. http://www.science.ku.dk/english/press/news/2015/planetary-boundaries/. Accessed 15 Jan 2015
Refugees (2018) UNHCR. https://www.unhcr.org/globaltrends2018/. Accessed 1 Feb 2019
Rich N (2019) Losing earth: the decade we could have stopped climate change. Picador, London
Robert K-H (2000) Tools and concepts for sustainable development: how do they relate to a general framework for sustainable development, and to each other? J Clean Prod 8:243–254
Schumacher EF (1993) Small is beautiful. Vintage, London
Senge P et al (2010) The necessary revolution. Nicholas Brealey Publishing, London
Science for Environment Policy (2015) European Commission. https://ec.europa.eu. Accessed 20 Oct 2021
Stockholm Resilience Centre (2021). https://www.stockholmresilience.org. Accessed Nov 2021
Sustainable Development Goals (2015) UN. http://www.un.org/sustainabledevelopment/sustainable-development-goals/. Accessed Dec 2015
The Brundtland Report (1987) WCED. https://sustainabledevelopment.un.org/content/documents/5987our-common-future.pdf. Accessed 9 Sept 2015
The Kyoto Protocol (2015) UNFCCC. https://unfccc.int/process-and-meetings/the-kyoto-protocol/what-is-the-kyoto-protocol/kyoto-protocol-targets-for-the-first-commitment-period. Accessed 10 Oct 2019
The Paris Agreement (2019) UNFCCC. https://unfccc.int/process-and-meetings/the-paris-agreement/the-paris-agreement. Accessed 3 Sept 2020

The Nine Planetary Boundaries (2019) Stockholm Resilience Centre. https://www.stockholmresilience.org/research/planetary-boundaries. Accessed 19 May 2019
UN Global Compact (2020). https://www.unglobalcompact.org/sdgs. Accessed 4 Dec 2020
United Nations. https://www.un.org/. Accessed 4 Feb 2021
Wallace-Wells D (2019) The uninhabitable earth: a story of the future. Allen Lane, UK
Yanarella EJ, Levine RS (1992) Does sustainable development lead to sustainability? Futures 24(8):759–774
6th Assessment Report (2022) IPCC. https://www.ipcc.ch/assessment-report/ar6/. Accessed 1 March 2022

Chapter 3
The Challenges of the Current Development Model

Abstract This chapter discusses the concept of development within the current economic framework. It deals with the pros and cons related to the economic growth trajectory founded on neoliberalism and capitalism that has advanced many nations globally, especially the rich nations, while the darker and destructive impacts have increased to the detriment of especially the poorer countries. This darker side that has led to crises related to climate, biosphere, and resources, subsequently causing a humanitarian crisis is examined. The chapter compares the current development model with the sustainable development model. Even though the word *development* figures in both models they are fundamentally different. The chapter defines these differences, and points to emerging symptoms that indicate a shift away from neoliberal capitalism brought about by the pandemic, and talks about the reasons behind this. The chapter discusses the role of the state as it is the entity solely responsible for the well-being, safety, and progress of its citizens. As such, it is the most vital factor in implementing sustainability. The chapter closes by looking at future perspectives and other factors to assess the possibility of implementing sustainable development in the current framework.

3.1 Development

> All critical approaches find development, as presently understood, to be a mistake of (natural and social) global proportions (Peet and Hartwick 2015, p. 311).

To address the issues that completed Chap. 2, regarding the status quo and the difficulty of making headway in trying to stop the global threats, we must look at why and how we have created this situation. We must face the facts. The nature of the type of development we have adhered to for 200 years or moreparticularly since World War 2, has a dark side. In the beginning, we observed the massive positive changes to our lifestyles in most of the Western world. Transformational material, scientific, economic, and profound social change have happened in my life so far. I cannot deny that. Neither can I deny having benefitted significantly in terms of health and well-being during that time. The unfortunate and disturbing aspect is that development did not benefit all of us. The poor became less poor but remained poor,

K. Blincoe, *Achieving Sustainability: The Ultimate Human Challenge*,
https://doi.org/10.1007/978-3-031-10023-9_3

and the social problems of many cultures were not solved. Women got voting rights but were, and are, still not regarded as equal to men and continued to be abused together with children, especially girls in many countries. Racism flourished. Even though the gap between rich and poor seemed to begin to close, the last few years, especially during the epidemic (Sect. 8.2), have seen the rich getting richer and the poor getting poorer. From being observers to benefitting directly as participants in the development frenzy, we have become responsible. By *we*, I mean particularly the people in the developed countries that now include most large economies on all continents. Here I do not distinguish between autocracies and democracies.

The Cambridge Dictionary tells us that *development* means "*the process in which someone or something grows or changes and becomes more advanced*". Other definitions echo this. Development can be defined as a process of transition from one state to another, whether related to human development or societal development. The use of *development* as a term in this book relates to the economic and political system that has reigned the world since World War 2. It is based on a combination of capitalist and neoliberal ideals. Some authors and researchers have defined development as a process by which nations improve living conditions for their citizens. This goal is found in most theories of development. They offer ideas on how to develop and improve societies at a national and international level (Piketty 2020). The belief was that the effect of the economic growth model would enhance the lives of all human beings. That was the theory. The practice turned out differently.

With the end of World War 2, a new era of development started, launched by American President Harry Truman in 1949. He said:

> We must embark on a bold new programme for making the benefits of our scientific advances and industrial progress for the improvement and growth of underdeveloped areas. The old imperialism - exploitation for foreign profit - has no place in our plans. What we envisage is a program of development based on the concepts of fair dealing (Esteva 1992, p. 6).

Another important word introduced by Truman was *underdeveloped*, which has become part of the development discourse since then. There are a few similarities between Truman's development goals and sustainable development. *Fair dealing* and *improvement and growth in underdeveloped* countries are similar. The most crucial significant difference is the environmental factor that is an integral part of sustainable development, and one that was not given any value in Truman's idea of development. The German author and researcher Wolfgang Sachs and the contributors to his book "The Development Dictionary" (1992) highlight the shortcomings of the traditional development goals and processes. Truman's development goals increased inequality rather than the opposite, and *exploitation for profit* continued. Much published evidence from a wide variety of sources has supported this point over many decades, such as the reports from UNICEF, UNDP, and the UN. Sachs stated: "The idea of development stands today like a ruin in the intellectual landscape. Its shadow obscures our vision". Today, many intellectuals, economists and theorists agree.

The Growth Predicament

It is not the aim of the book to discuss the broad field of development theories in depth but to comment on some of the adverse results of their implementation, as they are the reason sustainable development was launched as a new model in 1987. Some of the theories, for example, the modernization model (see below), succeeded in achieving better living conditions for many people worldwide. The published results of the Millennium Goals confirm this (Sect. 2.4). Despite the progress achieved, many problems started to emerge that have grown to now threaten the relatively stable climate patterns we have had for centuries. The focus of the capitalist development theories was narrow, economic and growth orientated without regard for keeping the underlying foundation for human life healthy and intact—a foundation that was and is both social and environment based. The environment did not figure in the advancement of industrialisation and globalisation. It was something to own, use, and exploit for the economic benefit of the few.

The Modernisation Model

The focus of the traditional approach was to develop society in terms of growth through industrial and technological advancement, generation of wealth, and global expansion. The way to achieve this included engaging in processes such as those described in the modernisation theory. It advocated a development process consisting of five stages, also known as "Rostow's stages of development" (Rostow 1960). The stages included a progression of society from initially being based on agriculture through the introduction of industrial growth, building a societal infrastructure, promotion of manufacturing, and advanced industrial and technological growth to the final stage of mass consumption, outsourcing of production and introduction of welfare initiatives. These stages would take six or more decades to complete. They formed the development trajectory that was followed by most Western countries and explain how these countries reached their high or advanced levels of progress. It was a process in which also third-world countries engaged. However, reaching the advanced stage (of economic growth) varied from country to country, and many developing or poorer countries were held back. The reasons for this are many and varied. They can to some degree be explained by a theory called the theory of dependence.

The Dependence Theory

The dependency theory elucidates how international capitalism maintained the status quo through, for example, *exploitative relationships* where the rich countries were being fed primary goods, raw materials, and other resources, as well as cheap labour by the poorer countries. However, these countries, in turn, became reliant on both the demand for labour and on imported goods from the wealthy nations. They became locked in a dependency trap (Peet and Hartwick 2015). This may sound simplistic, but although the dependency situation is complex, the above makes sense and reveals some of the reasons why many developing countries have struggled and still do to advance their societies. The dependency trap is still operating, mainly where mining is a significant industry and in countries where labour is cheap. It is leaving many developing countries vulnerable to external impacts, whether politically, economically, or environmentally.

Supporting the above is the fact that Western countries overall have benefitted most from development initiatives, whereas countries such as some on the African continent have benefitted the least and are caught in the dependency cycle. Studies show that approximately 15% of the world's population owns 80% of the world's wealth (Kingsbury et al. 2016). Poverty still exists. In 2016 around 1 in 10 people lived on a wage below $1.90 a day. 80% of these lived in just 20 countries. The evidence points to increasing inequality and a wider gap between rich and poor. This situation has grown during the pandemic, and we do not know as yet the entire trajectory of this. So even if Hans Rosling (2018) (Sect. 2.4) tells a positive story regarding global areas of social advance, the fact is that climate change is eroding these improvements and are leaving communities worldwide without the ability to sustain their basic needs. This is occurring while the rich get richer—some, even more during the pandemic. Inequality is a significant result of the unending economic growth obsession. Apart from climate change, inequality probably casts the second biggest shadow over the strive for and obsession with neoliberalism.

The Keynesian Model

The Keynesian model, is one of the more moderate and socially orientated development models or ideologies which ruled the political and economic scene, especially in Europe, after World War 2. John Maynard Keynes was an English economist who transformed Western economics from the 1930s to the 1970s. Keynes var a capitalist, but contrary to the neoliberal model, his ideas included socially orientated aspects and a belief in the state's role. The Keynesian regime advocated ideals such as high income, full employment, and free social services for everyone. Even though there was a belief in free markets, the state played a vital role as the moderator, implementer, and supervisor of these interventions. The model was overturned in the late seventies when the neoliberal ideals took over, picked up by Thatcher and Reagan, and implemented worldwide. It has been dominant in the Western world until today (Peet and Hartwick 2015).

3.2 The Current Paradigm

> We emphatically need a new political-economic theory to save the world, and it will not come from conventional economics (Peet and Hartwick 2015, p. viii).

The adherence to and belief in economic growth continued and grew stronger over time. Growing the economy and globalisation became the 'holy grail' of the post-war era, as the key lever of the capitalist system and the neoliberal ideology. This worship of monetary and material wealth has continued until now when a new agenda appears to be emerging from the problems that have accumulated and intensified in the last few decades, such as the devastations of a warming planet and the Covid-19 pandemic.

Capitalism

Capitalism is the foundation on which our current paradigm is based, interwoven with the neoliberal ideology ruling our politics and economics. We have adhered to the capitalist ideals for more than two centuries and neoliberalism for five decades. By 'we' I include most of the world, not only the Western hemisphere, as the majority of countries worldwide have adopted capitalist and liberal principles. Capitalism, sometimes also called the free market economy, has been dominant in the Western world since the eighteenth century, when it took over from the feudal system. Capitalist ideas have existed throughout history, but the theory the modern world has adhered to stems from the Scottish economist Adam Smith (1723–1790), whose ideas in the book "The Wealth of Nations" published in 1759 were regarded as the start of modern economics. Capitalism is an economic and political system where the means and profits of production are privately owned. Governments have little say. Private owners control production, and the aim is pure profit. Other characteristics of capitalism include the build-up of capital, production being dependent on supply and demand, division of labour, sharp competition, and adherence to private property rights. Little has changed since then in terms of privatisation of production and economic growth for profit.

Neoliberalism

Neoliberalism, as capitalism, stresses the importance of competition and the free market. It also promotes continued economic growth believing that this benefits and promotes human progress. It is more dynamic than conservatism and believes that change is good if it is the *right* change, whereas the conservatives believe in leaving things the way they are and protecting what is. The neoliberal project is both economic and political in nature. It can be seen as an extension of the capitalist system and focuses on the freedom of the individual, free market policies, growth of international trade, free flow of capital, deregulation of markets, reduction in taxes and public spending and encourages a flexible labour market. Supply and demand rule the global economy, not the governments.

Neoliberalism is an ideology and political model, developed by the Austrian economist Friedrich von Hayek (1899–1992). It was promoted particularly by the Mont Pelerin Society, which was an organisation set up in 1947 consisting of high-profile economists, historians, business leaders, and intellectuals who advocated and most effectively promoted the neoliberal model to governments around the world. Neoliberalism was picked up and introduced in the UK during the reign of UK Prime Minister Margaret Thatcher and in the US by American President Ronald Reagan, in the late 1970s. The ideology has spread and dominated the economic and political agenda until today. The two world leaders were very explicit in renouncing the Keynesian principles that had reigned until then. Margaret Thatcher declared that "there is no such thing as community, only the individual". President Ronald Reagan claimed in 1979 that governments were unable to solve crises, that the markets could, and that "government is the problem".

Although new ideas for change started to emerge after the last economic crisis in 2007–2008, the neoliberal ideals have remained largely intact (Peet and Hartwick 2015). The old ideologies lacked social and environmental values To address the current crises we must replace these ideologies and demand that the focus for development turns to nature-based and human potential, building resilience and repairing what has been lost or broken. The neoliberal ideas were fundamentally different from those of sustainable development. They cannot change the system and processes that are needed to move from unsustainable to sustainable practices. Business as usual is therefore not an option. The sooner we can re-imagine the better world that we know is possible, the sooner we can inspire to enact it and develop the methods and tools necessary, rather than staying with the dystopian scenarios that are manifesting around us.

3.3 Development Crises

The crises of development started many decades ago and were, of course, the reason why the Brundtland Committee was set up in 1985 that produced the report, "Our Common Future". The Committee responded to the problematic impacts of the development and globalisation period that were emerging in all areas of the world and started to affect peoples' lives. "Our Common Future" launched an appeal to governments around the world to create a process of a new development trajectory called sustainable development. Sustainable development was the response and remedy with which to address the current crises. The progress we had obtained placed us in a position where we could start to decrease and minimize the loss and damage to those who suffered and implement long-term adjustment initiatives. There was no longer any excuse for the loss of biodiversity, or global warming.

It is important in this context to briefly examine the crises, as they encompass more than climate change. They also include the crises of biodiversity, pollution, depletion of resources, and an ensuing humanitarian crisis. So much has been written about the changing climate. What we experience now was predicted many decades ago. For more than fifty years, research warned the world's governments that climate change could destroy societies and the way of life many had come to take for granted. Much research has been done, and many books have been written about the dangers of global warming. When researching for my PhD between 2015–2019 until today, the amount of literature on apocalypse, armageddon-like scenarios, and bleak prospects for the future of the planet and humanity has abounded—incredibly depressing reading. In the face of this, many people have become numb and feel powerless, turning to their small rectangular, digital screens, taking drugs, drinking too much, or engaging in other obsessive behaviours. Others have turned to mass grief rituals, become enraged, and have created organisations like Extinction Rebellion (2022) and have staged various demonstrations.

The IPCC 6th Assessment Report from 2021 stated the following regarding the current state of the climate (paraphrased):

- That the atmosphere, ocean, and land have warmed due to human influence and extensive changes have occurred.
- That these changes are unprecedented "over many centuries to many thousands of years".
- That human-induced climate change is causing extreme weather patterns worldwide, including droughts, heatwaves, storms and cyclones, torrential rain, and flooding.

What will occur in the future, if severe measures to re-address the situation are not taken rapidly, will be a surface temperature that continues to rise. Global warming will go beyond 1.5 and 2 °C and weather patterns will become more extreme. The Arctic Sea ice will continue to reduce and seas globally will warm further. The latest Assessment (2022) report goes further and gives us three years to break the curve of rising CO_2 emissions.

It is almost unreal writing this about the future, knowing that the necessary interventions have not and are not being initiated despite the overwhelming evidence. We all know that the remedy entails limiting CO_2 emissions to net zero. We also know it can be done. The United Nation's Conferences of the Parties, the COPS (Sect. 2.4), have been held bi-annually in various places around the world to address the crises, each attended by representatives of the world's nations. They have outlined many good intentions and produced many promises since 1992, which have never been met. When Greta Thunberg, at the Youth4Climate summit in Milan in Italy in September 2021, said the words: "Build Back Better. Blah, blah, blah. Green economy. Blah blah blah. Net zero by 2050. Blah, blah, blah," I understood what she meant.

The Environmental Crisis

Integrated into the climate crisis are other environmental problems that have continued to intensify. They include degradation of the environment and the natural habitat, biodiversity loss due to exploitation and pollution. Deforestation and desertification. Acidification and pollution of the seas. The state of the nine planetary boundaries developed by the Stockholm Resilience Centre that I mentioned in Chap. 2 illustrates this loss. These boundaries consist of nine processes that regulate the stability and resilience of the Earth's system. Eroding of soil is happening due to mass agricultural cultivation processes as well as desertification due to severe and long-lasting droughts. Increased focus is put on the problem of toxic and overfished seas as well as on the chemical toxicity of our environment in general, as the number of chemicals used seems to be multiplying rather than decreasing. The pollution of the seas is building up, the coral reefs are dying, and many species of fish are becoming extinct. The assessment reports from the IPCC, the Stern Review on the Economics of Climate Change (2007), the Stockholm Resilience Centre, and researchers around the world have all produced scientific data on this state of play. The many discussions and debates relating to the degree to which they have been caused by human activity or not are no longer valid as the reports have underlined and stressed the human factor in causing the impacts. I believe that everyone today will agree that there is sufficient evidence to alleviate any doubt that we have an environmental crisis, that it is growing day by day, and that we are responsible. Climate change is an integral part of this mix, and although it has taken precedence over other environmental problems, it is the combined effect of all the interconnected factors that, in time, will exacerbate the impacts. It is the domino effect that gives the greatest cause for worry.

The Resource Crisis

Another problem, which related to resources, became apparent in the late seventies as a result of consumerism and globalisation. It was due to the continuing increase in demand for resources including fossil fuels, metals and minerals generated from non-renewable sources both in developed countries but increasingly in developing countries. Resource loss and degradation became a problem due to overconsumption but also because post-production resources were mainly being discarded rather than reused, recycled, and upcycled. It was the beginning of a crisis that has deepened and will continue to deepen in years to come. The increase in resource use corresponded with the surging consumerism brought about by an expanding world population and higher standards of living. It has been estimated that more than three billion people from developing countries will receive middle-class incomes by 2050 (Kharas 2017), intensifying the resource need over the next decades. A different problem arose during the Covid-19 pandemic, where the virus affected everyone, also those in the supply industries, causing a backlog of containers of supplies that didn't reach their destinations. Supplies ceased, especially in the building and production industries. The war between Russia and Ukraine presents an even bigger problem regarding resources, which also encompasses food supplies. Due to the restrictions and sanctions, the world will increasingly suffer from a lack of a variety of resources, from fossil fuels to metals, minerals, wheat, and related food products. It is as though we are being forced by the virus and the war to rethink, reconsider, replan and redesign the way we live and work.

The resource issue has led to wars and civil unrest in Africa and the Middle East post World War 2 due to the fight to control non-renewable resources including minerals and oil. In some countries, wars and civil unrest became part of daily life, killing people, destroying livelihoods, cities, towns and villages and leaving devastation in their wake. The latter years of extreme weather added to this, causing an increase in numbers of refugees and migrants seeking new places to live and work. The author and lecturer Gwynne Dyer predicted in his book "Climate Wars" (2011) a future ruled by politics and wars for survival, as food stocks dwindle, with fresh-water shortages and loss of arable land on the increase, deepened by more floods and rising sea-levels. It could very well happen, as the estimated timespan to mitigate the effects of global warming is narrowing, and the impacts are escalating. The world could become a victim of its growth-based development agenda at a much larger scale than most people can imagine (IPCC 2020; Wallace-Wills 2019).

The Human Crisis

A human crisis emerged, intrinsically linked to the above. The African continent has been the primary victim of the Western development model as it, for example, contributed to the formation of an elite that owns most of the land and resources. This elite is seen to have undermined the self-sufficiency of local communities and has created an underclass of people who are poor and struggle to survive. The underclass includes women whose working and living conditions are poor, even though they are often both breadwinners and homemakers.

I worked on an educational project in Mali, Africa, from 2010–2014. At one of our seminars in Selingue, a village in the country a two-hour drive from Bamako, the local men and women were discussing gender issues. The men owned the land and the women, who often were the main breadwinners running the families and managing small holdings such as fruit orchards, rented the land from them. At the seminar, the women complained that it was difficult for them to harvest the fruit at the right time as the machinery for harvesting was also owned by the men. This meant that they decided who to rent to and when and prioritised the smallholdings that were managed by men. The women complained that they were always the last to get the tools they needed and often couldn't harvest in time. It resonates with the equality problems that exist everywhere. Despite, and as mentioned above, because of "development" inequality still exists both regarding gender and as income.

Statistics tell us that many women and children globally suffered the most during the pandemic due to domestic violence and abuse. Depression has affected many, mainly the young, and is said to become a global condition, which in my opinion, is not only due to the pandemic but also due to the current problematic predicament. Climate change has affected many people and communities worldwide. Approximately 80 million people (UNHCR 2018) are currently migrating from increasingly severe weather impacts, war-ridden areas, and volatile economies. It is estimated that there will be around 125 million climate refugees by 2050. The IPCC reports that the refugee and migrant crises will deepen unless sustainable development with a strong focus on social and human aspects is implemented worldwide within the next decade.

The crises, problems, and issues discussed above clearly illustrate how interconnected everything is and how one problem causes another. There are no boundaries, no borders, and no walls to keep these problems out. They are spreading and intensifying. We are all impacted. We created the crises and are the only ones who can provide the solutions.

3.4 Role of the State

When discussing development and societal change it is necessary to revisit the role of the state (Chap. 2), as it is the state that is the guarantor for making sure that development initiatives benefit all citizens. The state controls public service activities, distribution of income, the flow of money, investment, and the labour force. The state engages in and controls foreign liaisons and can make and pass the necessary rules and regulations to provide public services. It collects taxes and spends them according to agreed goals. Over the years, these roles have expanded to include the protection of property rights, ensuring a stable economy and today, also protecting the environment (Anter 2014). The state must, in short, be able to do what is needed to look after and safeguard its citizens and their environment. The belief in economic growth was anchored in the idea that a strong and developing economy would have a positive and qualitative impact on humanity in general. This trajectory made a good and convincing story. What was not thought through or explored was a projection of the negative aspects of the growth paradigm. Applying the precautionary principle early on would have avoided some of the emerging crises. The precautionary principle is a guideline based on ethics that states that no change should be introduced that might harm unless and until there is proof to the contrary.

As it is the key role of the state to protect its citizens, whether the threats are diseases, climate change or civil unrest and wars, it is interesting to observe the different approaches governments have taken in relation to addressing the threats. The first two are similar in certain respects. They are global, universal, affect everyone and are severe and life-threatening. The last two have frequently not been global and have been contained locally, apart from the world wars. However, they have had indirect and after-effects that impact us all. Concerning the virus, most governments took action immediately, implemented the necessary distancing and lockdown rules, opened the state coffers, and found the necessary financing to alleviate and combat the virus, and prevent an economic melt-down. It took approximately two years for the virus to be controlled globally and it was achieved solely through government intervention.

Regarding climate change the story is different. Governments have, since the Rio Convention, agreed to support each other as well as help the less advanced nations to combat the impacts of loss and damage and develop the necessary technologies to reduce global warming. Nonetheless, the commitments to date have not been sufficient to tackle the impacts of CO_2 emissions, which are still rising. None of the agreements, apart from the Paris Agreement in 2015, have been legally binding, and it has been left to the individual nation to decide what, how and when to act. The recent war has made the Western governments rally and offer military and humanitarian aid to Ukraine, apply extensive sanctions to Russia and have made the decision to militarise at extraordinary costs.

It seems then that states lack the capacity to deal with long-term planning in the context of adapting to and mitigating threats such as climate change or engaging in sustainable development. Some say that states are ill-designed for such planning. I don't believe this is so. When securing resources for future production and consumption, we have seen states engage in detailing long-term planning strategies that have involved wars and the manipulation of international relations. If it is a key responsibility of the state to protect its citizens from crises such as those described in the section above, the state has failed both regarding its short as well as long-term responsibilities related to the climate impact. It is a failure that today threatens the survival of human and other living species, and for which the state should be held accountable. It is not that the state is ill-designed; instead, I believe that it is the people who manage the state that are at fault. Their focus is on short-term plans and gains, such as pleasing the electorate and staying in power. They seem to be influenced by powerful global actors, especially in an economic context who are willing to sacrifice the future to benefit the present.

3.5 The Sustainable Development Model

> Development that meets the needs of the present without compromising the ability of future generations to meet their own needs (UN 2015).

United Nations took the initiative to address the crises as the representative of the world's nations—and so, the Brundtland Commission founded by the UN in 1983, launched the concept of sustainable development in 1987 (Sect. 2.2). The wide issues to be addressed represented a complex challenge given the current paradigm into which these concepts were introduced. As the classic quote above illustrates, sustainable development aims to achieve well-being for humanity within the carrying capacity of the earth. This aim to increase human well-being is one factor that ties it to other development theories. Where it differs from the conventional theories is its focus on respecting, protecting, restoring the natural habitat, and reversing the adverse impacts. It advocates promoting social equity and achieving global equality in an economic context.

Revisiting the Brundtland recommendations I have chosen several elements that are key to its realisation including the following:

- A social system that provides for solutions to the tensions arising from development.
- A political system that secures effective citizen participation in decision-making.
- An economic system that is able to generate surpluses and technical knowledge on a self-reliant and sustained basis.
- A production system that respects the obligations to preserve the ecological base.
- A technological system that can search continuously for new solutions.
- An international system that fosters sustainable patterns of trade and finance.
- An administrative system that is flexible and has the capacity for self-correction (WCED 1987).

The importance of taking a systemic approach to implementation of sustainable development cannot be underestimated. Although we have seen uncountable and remarkable pioneering initiatives worldwide since 1987, the systemic intervention that is needed is still lacking. Some of the above recommendations have happened, such as the continued technological advance and the search for new solutions. Most governments are relying on not yet invented technological ways and methods to combat global warming.

The most radical diversion from traditional development to the sustainable development path is two-fold: The first is the focus on the natural habitat, including the urgent and needed intervention to maintain, protect and restore nature, maintain biological diversity, replenish and maintain the natural ecosystems, regeneration, and re-wilding. Stopping the overuse of natural resources. This focus reflects the recognition that a healthy planet is a precondition for the survival of the human species—we cannot live on digital food or algorithms. The second focus is on social well-being and cohesion. As previously mentioned, there is still too much poverty, hunger, lack of quality education, adequate healthcare, gender equality, income equality, and many other socially orientated problems globally. These key focal points must be addressed as this is where the current paradigm has failed. There has not been any focus on the natural habitat nor an assessment of its intrinsic and economic value even if conventional development theories have aimed for the well-being of citizens and societies. The above list of lacks, points in the opposite direction. Traditional theories set in the current neoliberal paradigm and measured by GDP and DNP cannot meet the challenges of a transition to a state of sustainability. The essential nature of the current belief in economic growth is based on materialism and is production, consumption, and profit-orientated.

There has been a variety of approaches to sustainable development put forward by many researchers. John Blewitt, who in my opinion, is a critical and relevant researcher and writer on the topic, mentions several interpretive views in his book "Understanding Sustainable Development" (2008). He points out that there is a request for "rigorous concepts and key principles focussing on self-restraint, balance and a spiritual dimension" and further on "the concept of political ideology that best links ecological themes with social struggles" and "processes of dematerialization" and "resocialization", which uncouple economic activity from ecological impact". Other researchers also stress the importance of a cultural approach. These ideas reflect the dichotomy, multidimensional and multidisciplinary aspects of sustainable development. It must have been relatively straightforward to implement a development model that focussed on economic growth while believing that societies would benefit as a natural consequence of this pursuit. Given that the neoliberal model began its well-promoted implementation process in the 1980s and spread globally within a few decades, it should concern us that the model of sustainable development will take many more decades to be fully implemented—if at all, as it has been met with overwhelming resistance since 1987.

The 17 Sustainable Development Goals

Many researchers, economists, engineers, architects, and a variety of other experts have attempted to develop and apply methods to engage in sustainable development activities. The more well-known initiatives are discussed in Chap. 5. The latest model, the 17 Sustainable Development Goals, was developed by the UN as an endeavour to explain, define and break down the concept of sustainable development into realisable goals or units. Each of the 17 goals can be addressed and worked on by all actors of society. The goals include subgoals and indicators to help the process. A time frame has been given, initially 2030, as a milestone, although it is unrealistic. However, the 17 SDGs represent a framework or a global strategy for sustainable development. It is the only one in existence that covers such a broad spectrum of issues or topics, and as such, it is the best we have.

The 17 Sustainable Development Goals were agreed upon in September 2015 by 193 countries at UN Sustainable development Summit in New York. The 17 goals built on decades of initiatives undertaken by various countries in collaboration with the UN. They were a replacement and an extension of the Millennium Development Goals (MDG) from 2000. The Millennium Goals were also developed by the UN and consisted of eight goals to be met by 2015. They focused on alleviating problems such as poverty, hunger, education, gender equality, and health issues relating to mortality and diseases, environmental degradation, and strengthening global partnerships. The Millennium Development Goals were to be achieved specifically in and for developing countries supported and financed by the rich nations (see further description of the 17SDGs and the Millennium Goals in Sect. 2.3).

The eight MDG focus areas were repeated in the new set of goals with the addition of another nine topics such as clean water and sanitation, biodiversity, renewable energy, climate change, inequality, sustainable cities, and communities, improved working conditions, and peaceful institutions (UN 2000). The new set of goals differs from the Millennium Goals by the fact that they relate to and must be achieved by all nations, both developed and developing countries, and that they include a wide spectrum of focus areas.

The 17 SDGs were created by a cross-section of world actors: the UN, governments, the business sector, and civil society. The goals are universal, interdependent, and transformative. Achieving them requires all sectors of society and all nations to work together in multi-stakeholder partnerships to share knowledge and technologies, expertise, and financial support—as well as trials and errors (UN 2020). With this new segmented framework, it has become possible to work with goals that are relevant to specific areas of work. The sectors that have engaged with the goals, have, in the main, firstly chosen to work on problems that can be easily addressed and changed before moving on to tackling more complex issues. The 17 SDGs allow for this. The interrelatedness of the goals means that focussing on one will affect other related areas—the aim is to subsequently reach all goals, thus achieving a state of sustainability.

In Denmark, many actors have engaged with the 17 SDGs since 2015. Municipalities, cities, communities, schools, businesses, organisations, and NGOs have committed to work with the goals. A cross-section of politicians from all parties are members of a network called the 2030 Panel, where they meet regularly and have invited a selection of people from civil society for discussions in Parliament on implementation potential and perspectives. Recent work, for example, focused on how to implement Goal Nr. 12 which deals with ensuring sustainable consumption and production patterns. The project included obtaining the relevant statistics and information to develop appropriate scenarios, processes, and methods to reach the goal, using indicators based on data relevant to Denmark. The emerging challenge that this project represented was the need for accurate statistics, information, and facts necessary to develop the right indicators to monitor and measure the results (Hildebrandt 2018). This initiative has highlighted the need for accurate data, without which no appropriate action can be taken. Another challenge that has emerged relates to the culture and tradition of each country, as the goals need to be adapted to include and embrace the cultural differences. Danish consumption patterns differ from those in other countries. When documenting the barriers to sustainability in Chap. 7, the cultural barriers play a significant role in resisting the transition process.

Opinions regarding the potential or likelihood of achieving the goals by 2030 are divided. When the goals were introduced in Denmark, many actors felt relief due to the break-down of the sustainability construct into 17 focal points that contained both sub goals as well as indicators. Novozymes (Sect. 6.8) adopted the 17 SDGs and made them the framework for their future strategy, as have many similar Danish companies. However, as time has passed, the initial enthusiasm has receded to some degree. Addressing and meeting 17 goals is a challenging task, especially as it is a new model that has neither been tried nor tested. When the initial enthusiasm vaned, some actors started to turn towards climate change—Goal Nr. 13—and many have now made this their exclusive focus. Understandably so. CO_2 reduction is an immediate need, it has global attention, it is a specific challenge, and it is generally seen as a problem to which there is both a technological solution and a personal one. For example, less consumption, less waste, less meat, more electric vehicles, more public transport, more heat pumps, more recycling, and so on.

Some examples of *greenwashing are* associated with the 17 development goals. As the process of implementation met the complexity barrier and other obstacles, the easy way out was to pretend to be working with the goals. Many websites support this view. Low-hanging fruits became the focus, which consisted of existing elements that businesses, organisations and even governments promoted as their new contribution to lowering CO_2 emissions. It is, therefore, difficult to statistically assess how far the work with the SDGs has been taken. Research and books have yet to be published on the effect of the goals and the application results. It is now some years since they were launched, and more time may be needed before any significant effect can be measured or valid judgements be made. Meanwhile, CO_2 emissions are still rising.

3.6 Perspectives

> Real change requires the truthful acceptance of "what is" (Harvey 2019, p. 13).

I believe that deep change is on the horizon. Some claim that neoliberalism *died* during the pandemic. This time the market forces could not solve the crisis. Governments took control of their nations and became all-powerful through interventions to curb the virus and keep their societies functioning through a massive public capital injection. The pandemic also showed the need for *us* rather than *me,* indicating that the era of focussing solely on the individual rather than on the community was over. Although the declaration on the end of neoliberalism can be called into question, as we still do not know the long-term effects of the pandemic, there are indications that a fundamental change is occurring. For example, the way of working has changed with people working from home if not five days of the week, then 2–4 days a week. This is becoming the new norm. In addition, there seems to be a new attitude to work. More resign from their jobs due to unsatisfactory working conditions or the sense that work is meaningless. The #MeToo movement is also here to stay, indicating that a new work culture and ethics will become prevalent—long overdue. Still, it is here (see also Sect. 8.2). Another shift was the sudden separateness or compartmentalisation of our societies due to the virus. This shift introduced the creation of visible and invisible borders, where people were withdrawing, becoming introverted, building physical barriers, and rekindling the importance of smaller groups. Borders became increasingly accepted to keep the virus out. We started to retract and refocus our attention on the here and now including our relation to the local environment, close families, and friends. In many countries, governments have regained the respect and trust of their citizens, who obeyed and followed the many devastating rules of lockdown during the pandemic. The retraction and giving up personal freedom for the benefit of the collective and the national *tribe* have become the new way and I believe, it will have a lasting impact on most of us.

Other factors that might help to replace neoliberalism with another economic model are the current impending crises of climate, environmental, resource, and biodiversity degradation. We must work together to deal with the challenges, so the collective becomes more important than the individual, and sacrifices must be made. We will not go back to the Keneysian model as the number of diverse and different factors calls for a new kind of economic and political agenda. The new agenda will be based on collaboration across all types of divides, from the personal to the public, from local to global. At the same time, individual freedom will be curbed in terms of consumerism and lifestyle. The individual must step back for the benefit of the whole. The responsibility for this to happen belongs to the state with support from its citizens. Without laws, rules, and legislation, CO_2 emissions, for example, cannot be curbed. As to which new model will fit the future is an interesting question. Most literature criticises the existing system. Some new economic theories are, however, emerging, such as those described in the economists' Kate Raworth's "Doughnut Economics" (2017), Thomas Piketty's "Capital in the 21st Century"

(2013, 2019), and Tim Jackson's "Post Growth" (2021). They all advocate changing the economic system and talk about a post growth scenario where other more human and environmental values are central to a new economic ideology.

Kate Raworth's model, for example, is doughnut-shaped, where human prosperity is contained in a *safe* central space that is environmentally and socially safe as well as just. The safe space is bordered by an *ecological ceiling* as the outer edge, beyond which are the environmental problems we are facing today. It has an inner *social foundation* ring that includes twelve basics of life that every human being should have access to, such as food, water, sanitation, education, and decent work, identical to the 17 Sustainability Goals. The doughnut is thought of as a *compass* to guide our future direction. Kate Raworth's model is gaining recognition by many global actors such as cities. In a lecture at Copenhagen University on 19 April 2022, she said that 40 cities have adopted the doughnut model and are working to realise its potential (Copenhagen University 2022). Very encouraging.

Apart from alternative economic models, other theories on development of societies have emerged throughout time. As sustainable development encompasses much more than economics, I investigated the field of alternative and visionary theories to find models with a closer link and similarity to the construct of sustainability. The next chapter discusses this field.

Readings and References

Agyeman J, Bullard RD, Evans B (eds) (2003) Just sustainabilities: development in an unequal world. MIT Press, Cambridge US
Anderson RC (1998) Midcourse correction. Pereginzilla Press, Atlanta
Anter A (2014) Max Weber's theory of the modern state. Palgrave Macmillan, UK
Blewitt J (2008) Understanding sustainable development. Earthscan, London
Blincoe KS (2020) Essential sustainability: from utopian theory to practical application. Dissertation, University of Kingston. https://eprints.kingston.ac.uk/id/eprint/51146. Accessed 2020–2022
Brannen P (2017) The ends of the world. Oneworld Publications, London
Brown LR, Gardner G, Halweil B (1999) Beyond Malthus. The Worldwatch Institute, New York
Brundtland Report, Our Common Future (1987) WCED UN. https://sustainabledevelopment.un.org/content/documents/5987our-common-future.pdf. Accessed 10 Jan 2021
Carney M (2021) Value(s): building a better world for all. William Collins, Croydon
Crutzen P, Stoermer EF (2000) The Anthropocene. IGBP's Glob Change Newslett (41)
Daly HE (1996) Beyond growth: the economics of sustainable development. Beacon Press, Boston
Diamond J (2011) Collapse: how societies choose to fail or succeed, 2nd edn. Viking Penguin, New York
Doppelt B (2003) Overcoming the seven sustainability blunders. Syst Thinker ® 2(14):5
Doughnut Economics (2022). Copenhagen University. https://www.sustainability.ku.dk. Accessed 23 April 2022
Dyer G (2011) Climate wars: the fight for survival as the world overheats, 2nd edn. Oneworld Publications, Oxford
Ekins P (1992) A new world order. Routledge, London
Esteva G (1992) Development. In: Sachs W (ed) The development dictionary. Zed Books Ltd., London
Exstinction Rebellion (2022) Rebellion. https://rebellion.global/. Accessed 19 April 2022

Figueres C, Rivett-Carnac T (2020) The future we choose. Manilla Press, London
Goldin I (2018) Development. Oxford University Press, Oxford
Harvey M (2019) Utopia in the Anthropocene: a change plan for a sustainable and equitable world. Routledge, Abingdon
Hawken P, Lovins, Lovins AB (1999) Natural capitalism: the next industrial revolution. Earthscan, London
Heinberg R (2011) The end of growth. Clairview Books, Sussex
Hildebrandt S (2016) Bæredygtig Global Udvikling. Jurist-og Økonomforbundets Forlag, København
Hildebrandt S (2018) Verdensmålene - det vigtigste punkt på dagsordenen. KLS Pureprint, København
Hopper P (2012) Understanding development. Polity Press, Cambridge
IPCC Reports (2020) IPCC. https://IPCC.ch/reports/. Accessed 20 April 2020.
Jacobs M, Laybourn-Langton L (2018) Forum: a new economic paradigm. Paradigm Shifts Econ Theory Policy 53(3):113–118
Jackson T (2021) Post growth: life after capitalism. Polity Press, Cambridge UK. https://www.intereconomics.eu/contents/year/2018/number/3/article/paradigm-shifts-in-economic-theory-and-policy.html. Accessed 7 May 2020
Kharas H (2017) The unprecedented expansion of the global middle class an update. Global Economy & Development Working Paper (100)
Kibert J et al (2012) Working toward sustainability: ethical decision making in a technological world. Wiley, Hoboken
Kingsbury D et al (2016) International development: issues and challenges, 3rd edn. Palgrave, London
Kirby J, O'Keefe P, Timberlake L (eds) (1995) The earthscan reader in sustainable development. Earthscan Publications Ltd., London
Klein N (2014) This changes everything. Simon & Schuster, US
Limits to Growth (2019) Club of Rome. https://www.clubofrome.org/news/publication-of-the-limits-to-growth/. Accessed 10 Oct 2019
Meadows DH, Randers J, Meadows D (2004) Limits to growth: the 30-year update. Chelsea Green Publishing Company, White River Junction
Millennium Goals (2000) UN. https://www.who.int/topics/millennium_development_goals/about/en/. Accessed 6 Aug 2018
Peet R, Hartwick E (2015) Theories of development: contentions, arguments, alternatives. The Guildford Press, New York, p viii
Piketty T (2013) Capital in the twenty-first century. The Belknap Press of Harward University Press, Cambridge, US
Piketty T (2020) Capital and ideology. Harvard University Press, Cambridge
Planetary Boundaries (2015) Copenhagen University. http://www.science.ku.dk/english/press/news/2015/planetary boundaries/. Accessed 15 Jan 2015
Porritt J (2005) Capitalism as if the world matters. Earthscan, London
Princen T, Maniates M, Conca K (eds) (2002) Confronting consumption. MIT Press, Cambridge
Raworth K (2017) Doughnut economics: seven ways to think like a 21st-century economist. Random House Business Books, London
Refugees (2018) UNHCR. https://www.unhcr.org/globaltrends2018/. Accessed 1 Feb 2019
Rich N (2019) Losing earth: the decade we could have stopped climate change. Picador, London
Rio Declaration (1992) UN. https://www.un.org/en/development/desa/population/migration/generalassembly/docs/globalcompact/A_CONF.151_26_Vol.I_Declaration.pdf. Accessed 5 Feb 2019
Robert K-H (2000) Tools and concepts for sustainable development: how do they relate to a general framework for sustainable development, and to each other? J Clean Prod 8:243–254
Rosling H, Rosling O, Ronnlund AR (2018) Factfulness. Sceptre, London
Rostow (1960) The stages of economic growth: a non-communist manifesto. Cambridge University Press, Cambridge

Sachs W (ed) (1992) The development dictionary: guide to knowledge as power, 5th edn. Zed Books Ltd., London
Senge P et al (2010) The necessary revolution. Nicholas Brealey Publishing, London
Smith A, Darnell T (ed) (2018) The wealth of nations. 12th Media Services
Stern NH (2007) The economics of climate change: the stern review. Cambridge University Press, Cambridge, UK, p 2007
Sustainable Development Goals (2015) UN. http://www.un.org/sustainabledevelopment/sustainable-development-goals/. Accessed Dec 2015
The Brundtland Report (1987) WCED. https://sustainabledevelopment.un.org/content/documents/5987our-common-future.pdf. Accessed 9 Sept 2015
The Paris Agreement (2019) UNFCCC. https://unfccc.int/process-and-meetings/the-paris-agreement/the-paris-agreement. Accessed 3 Sept 2020
The Planetary Boundaries (2019) Stockholm Resilience Centre. https://www.stockholmresilience.org/research/planetary-boundaries. Accessed 19 May 2019
UNDP (2020) UN. https://www.undp.org
UNICEF (2020) UN. https://www.unicef.org
Wallace-Wells D (2019) The uninhabitable earth: a story of the future. Allen Lane, UK
Yanarella EJ, Levine RS (1992) Does sustainable development lead to sustainability? Futures 24(8):759–774
6th Assessment Report (2022) IPCC. https://www.ipcc.ch/assessment-report/ar6/. Accessed 1 March 2022

Chapter 4
Embracing and Learning from Utopian Thought

Abstract This chapter examines alternatives to the current growth model and delves into utopian thought and constructs. It offers a look into history, from Plato's the 'just' man and the 'just' society to the present time's small intentional community, and ecovillage ideals and examines the impact of both utopian and dystopian thinking on societies. Utopia is relevant in this context due to the fact that sustainable development is a radically different societal model to the one currently in place. Utopian ideas are radical, and whilst conventional theories of development may have the same aim to improve the lives of citizens, it's from utopian theories that we can learn about the nature of radical models, and how they have succeeded or failed in impacting societies over time. Linking this to the sustainability construct and the attempts at translating it into practical application, utopia provides contexts and knowledge to learn from, including the barriers that are preventing it from being realised. The chapter looks at real-world utopian experiments, where some have failed, and some still exist, which provides additional practical knowledge. The enquiry into utopia spans the period from the time of Plato to the present.

4.1 Searching for Utopia

> The power of utopian thinking, properly conceived as a vision of a new society that questions all the presuppositions of the present-day society, is its inherent ability to see the future in terms of radically new forms and values (Bookchin 1980).

The sustainable development model, as discussed in Chap. 3, places an equal emphasis on the natural habitat and humanity. An emphasis not seen in any adopted political or economic development theory until now. Yet throughout history and especially during the nineteenth and twentieth centuries, visionary models for alternative societies emerged that included nature as part of the equation and that emphasized its intrinsic value. However, the models were individual utopian explorations and experimentations. Most were meant for reflection and inspiration and not necessarily for implementation, but many became experiments, and were realised. Another of the main conclusions drawn from the chapter on sustainability, is the conundrum that despite attempts at addressing the urgent problems, the necessary interventions to

K. Blincoe, *Achieving Sustainability: The Ultimate Human Challenge*,
https://doi.org/10.1007/978-3-031-10023-9_4

combat and alleviate them have not been made. The most obvious reason is the inability or unwillingness by governments, and the business sector, in particular, to embark on the profound changes that are needed. Instead, there are attempts to reform society through individual and fragmented initiatives whilst doing business as usual. More transformative interventions are called for, and this, of course, is holding back the transition. A broader context and understanding are required to move forward. Contributions to this can be found in history, in particular in the field of utopian thought. The chapter therefore investigates and searches for similar challenges in a historical context. I believe there is a strong link between the construct of sustainability and utopia—more so than between sustainability and the conventional development models. My PhD research unveiled several similarities between the two concepts, which this chapter will illustrate.

The quote on utopia introducing the chapter expresses the opinions of many researchers and writers who have discussed and promoted constructive and positive aspects of utopian thinking, including its merit and importance in the development of societies. The quote confirms the widely accepted notion that new concepts or ideas have throughout history been designed either as an opposition to the present, fear of the future or as an attempt to aspire to a better future—very similar to where we are today. A better future in a utopian context is usually related to wanting an improvement in human well-being. A parallel to the essence of sustainability, which also aims to increase well-being in a holistic sense for all living species. Achieving sustainability requires an approach that is non-conventional and radical. Michael Harvey, referenced in Chap. 2, states in his book, "Utopia in the Anthropocene" that: "utopias are always radical rather than reformist. They argue that what's needed is not just a modification of the existing system but a completely different system" (Harvey 2019, p. 6). The more I researched the topic of utopia the more I understood how important utopian thinking has been for the development of our societies due to the impact it has had in the way of inspiring, advocating, and bringing forth new ideas and experiments, to improve our living conditions.

The Nature of Utopia

> Utopias hold up a mirror to the fears and aspirations of the time in which they were written: In that sense utopianism is always in fashion (Goodwin 2001, p. 2).

Some authors believe that utopian ideas came about as a result of political change, or social developments, including transitions and periods such as the discovery of the New World, the Age of Enlightenment, and the French Revolution. Some believe that utopian ideas have had transformative powers, have brought about societal changes, and influenced decision-makers. Others argue that utopian ideas have rarely received any recognition by politicians and have merely been some kind of fantasies without any real purpose or meaning (Miles 2008). Publications from the twentieth century, such as George Orwell's classic book "1984" (1949), Aldous Huxley's "Brave New World" (1946), Kurt Vonnegut's "Player Piano" (1952) as well as Ernst Callenbach's "Ecotopia" (2009) contain imagined scenarios or thought experiments on how society might or could develop. In some instances, however, predicting dystopias rather than utopias.

Utopian concepts cover a long and broad range of ideas over time that spans from Plato's political "The Republic" ca. 360 BC. to Thomas More's "Utopia" (1516) and Jacque Fresco's The Venus Project (1990s). The consensus by utopian writers, whether in praise of utopia or being critical, is that utopianism still has a role to play in creating alternative visions, models, or scenarios. Some of the utopias I was most inspired by were created by writers such as Henry Thoreau, who wrote and lived his utopia of ultimate simplicity described in the book "Walden—or Life in The Woods" in the1850s, where Thoreau spent two years in the woods in a self-built hut living from own grown vegetables and products, spending his time observing, writing, and drawing his life with nature. Another is the Scottish William Morris's utopia of austere beauty, in the classic "News from Nowhere" from 1890. William Morris was an egalitarian and wanted to transform the relationship between humanity and nature where beauty should play a significant role. A later writer, Ernest Callenbach, describes in his book "Ecotopia" from 2009 a region/country that has adopted sustainability. Even if these visions of an imagined future appear far-fetched, apart from Henry Thoreau's, who lived his Walden, there is a need to acknowledge these visionaries and scholars, listen to their stories and narratives and study their scenarios, as we might learn something. The past contains clues to the future.

Paradise or State

Utopia can take the form of myths, political manifestos, or constitutions. Plato's "The Republic" (427–347 BC) is famous for being one of the earliest political utopias. Plato was a disciplinarian and authoritarian, and his republic should be ruled rigorously. Some of the ideas and principles included:

- Equality between men and women.
- The rulers or guardians (as they were called) must live modest lives.
- Private property is banned.
- Luxury living is looked down on.
- Frivolity, anti-social behaviour, and artistic activities are banned.

Where Plato's ideal society was based on a mixture of political and social principles, other utopias dealt solely with social issues, and the earliest models were based on religious beliefs relating to creating paradise on earth. Utopia was in early literature until the seventeenth century, depicted as a place or state which existed mainly in theory and was generally regarded as a form of paradise based on divinity. From that time, many intentional communities or settlements, mostly smaller in size, were established within specific frameworks based on a combination of religious, social, and political ideals and principles. Today thousands of intentional communities exist around the world founded on such visions. Some of the more recent examples from the twentieth century include environmental factors and are described as ecological utopias also called ecotopias or sustainable utopias.

4.2 Definition of Utopia

Utopia means no place, from Greek ou = not + topo = place. It has been translated into meaning *nowhere, a nowhere place.* Some of the other terms applied to describe utopia were *bliss, Eden, Elysium.* Another version of utopia is *eutopia*, meaning a good place. The concept developed and changed over the years in tandem with the advance of our societies, although the essence has remained the same (Claeys and Sargent 1999). Utopia, as a concept through literature usage, has come to mean the following:

- An ideally perfect state, especially in its social, political, and moral aspects.
- An imaginary or real place considered to be perfect or ideal.
- An idealistic scheme for social and political reform.
- An ideal or perfect place, or state of things.

Utopia can thus be a society, a community, a settlement, a commonwealth, a village, a town, a city, or a nation. What fundamentally characterises this type of imaginary model is that it is always set within a specific time and place context. It is an imagined but practical idea and a philosophical thought experiment (Goodwin 2001), which for many centuries remained an idea until pioneers took the concept and turned it into living experiments.

There are derivatives of utopia such as dystopia or negative utopia that signify the opposite, meaning a place that is worse than the existing. As utopia deals with seeking human well-being achieved through religious, social, or political means, dystopia expresses extreme and adverse conditions associated with human suffering. Much visual creativity has been invested in the designs of utopias and dystopias. Especially today, where dystopian scenarios in particular, are turned into entertainment by the film, digital and social media industries in the form of apocalyptic and futuristic doomsday films, TV series, or computer games. There has been a significant increase in the number of dystopias during the late twentieth and beginning of the twenty-first centuries and an equally significant reduction in the number of new utopian scenarios. Utopias (*the good places*) are rarely depicted today. Among the few exceptions from the twentieth century are "Lost Horizons"—a filmatized book about an oriental paradise by the English novelist James Hilton (1933/1937) and the more recent film "Avatar" produced by the film producer James Cameron in 2009. One of the reasons for the lack of utopian visions today is the fact that most people react to the term utopia, believing it to be an idea too far-fetched to realise. The term has become derogatory, unfortunately. I believe imagination is vital to overcome the stagnation of ideas and the general powerlessness people feel about the future.

The term utopia was manifested and made public by Sir Thomas More as far back as 1516. Thomas More was a lawyer, philosopher, author, humanist, and justice of the King's Bench under King Henry VIII. The book is about the island of Utopia the description of which is narrated by Hythloday, a Portuguese traveller and scholar who visits the island. Utopia is a beautiful island in a far-away place, where people are happy, contented, and live fulfilled lives. The citizens are working with what comes natural to them, and the societal framework and structure seem to work to everyone's satisfaction. The psychologist, Michael Harvey, published a book on contemporary utopia called "Utopia in the Anthropocene" in 2019, which I have referred to previously in this book. He was inspired by Thomas More and "Utopia" and compared the background and framework five hundred years ago to our current time, maintaining that both periods relate to pivotal historical moments as they signify the meeting point of two eras. "Utopia" was written at a time when the modern or the new world was meeting the old order. The same can be said of today, where we must choose between a utopian or a dystopian future. Harvey hopes that we design the future of the post Anthropocene era with "hindsight, insight and foresight", a sentiment I believe we can all support.

4.3 The Role of Utopian Thought

> ..utopias are always radical rather than reformist. They argue that what's needed is not just a modification of the existing system but a completely different system (Harvey 2019, p. 6).

Utopian thought fulfils various roles as mechanisms for release or escapes from reality and as a source of ideas for new horizons, as they bring encouragement and hope when people are faced with complex and impossible circumstances in their daily lives and environments. Utopia provides a safety valve or a way to dream of another possibility.

The scholar and now Professor Emeritus in history at the University of London, Gregory Claeys, who I cite above, stresses the importance of these alternative ideas in his book of utopian ideas, "Searching for Utopia" from 2011. He infers that the human being is continually on a journey to improve his or her living conditions, changing what is not working and suggesting methods, principles, and ways forward. Without utopia there would not have been the impetus to better people's lives to the degree that has seen a continuous push of boundaries to improve our circumstances.

Each period has its utopia and ideas for the future, and we are now urgently having to define and implement a new model. The American developmental biologist and writer Dr Bruce H. Lipton maintains that the civilization, we are currently part of, is dissolving, just as other civilizations have emerged, risen, and fallen in the past. It is a similar view to Jared Diamond's account of earlier civilizations that have disappeared. However, this time, in the Anthropocene, we have the knowledge, technology, economy and know-how to make a conscious choice concerning the design of the next wave, the next paradigm. The evolved and wise choice would be a sustainability paradigm. Nothing else makes sense.

A History of Ideas

Utopianism is, by its supporters, regarded as a beacon guiding us towards the future. From my perspective, utopianism is about imagining and envisioning the future. Whether this envisioned future is implementable is not important as the value lies in the intention to strive towards something positive. The American historian, sociologist and philosopher Lewis Mumford sees in his book, "The Story of Utopia" from 1922 the world of utopias divided into two main groups of "escape" and "reconstruction". He views utopianism to be crucial, as it is human nature to want to continually look ahead and work towards the improvement of one's life and environment.

On the other hand, there are critics of utopian thought, who believe utopia to be unrealistic and impractical bordering on the naïve. Some view these visions as static and even dangerous. To support the critics, it is true that totalitarianism, symbolized by fascist and communist dictatorships, emerged from utopianism. The ideals were initially thought of as visions of better societies and thrived for perfection. However, instead, they advocated control and obeyance of the leaders to the detriment of the people. Utopian ideals such as these reveal the dark side of utopianism and demonstrate how a utopian model can change into dystopia through a chain of events. Events that were not resisted but encouraged often by a small group of influential people led by demagogues. The uncanny aspect of this is its similarity to the neoliberal political and economic model ruling the world today. Despite the harmful and dangerous effects of the model on many people around the world, as well as on nature, a powerfulelite still vigorously resists any attempts to course correct. What the eighteenth-century Scottish economist Adam Smith (1723–1790), the godfather of the modern theory of capitalism, thought was a model that would benefit all people turned into a dystopia threatening the long-term survival of humanity.

Utopia and Dystopia in Recent History

The gap between utopian theory and practice seems as wide as the gap between the theory and practice of sustainability. Many imaginative dream worlds and alternative societies have been developed as theoretical constructs. Some of these have subsequently inspired and influenced the thinking of people who have wanted to instigate change and have done so through experiments to realise utopian ideas.

The German sociologist Karl Mannheim, who wrote the book "Ideology and Utopia" (1936), states that the ideas of utopia changed during the twentieth century. New movements emerged, and he predicted two strands of future utopianism. One was based on old socialist ideals, which he believed would only re-emerge if an underclass of people was created. The other strand was the rise of the intellectual classes and their visions, which would see new and different scenarios emerge. The professor of politics at the University of East Anglia, Barbara Goodwin and her colleague, Keith Taylor, picked up on the predictions set forth by Mannheim in their 'The Politics of Utopia' published in 2009. They viewed the new intellectual movements as a response to the political changes globally, combined with the development of technology and the rise of environmental threats. Several other movements emerged. Amongst those were fascism and Marxism, technocracy, feminism, and communitarianism. The first two ideologies rose and disappeared as political models within a relatively short time period. However, fascism still exists and appears from time to time in extreme right-wing groups and networks. The other three are still present and growing today. Feminism is becoming stronger and is supported by focused attempts by governments, especially in Northern Europe and Africa. More and more women are being included in governments, as leaders, researchers, on company boards and in politics. The #MeToo movement, which is spreading globally, has enhanced and promoted feminism. It is absurd that women in 2022 are still regarded as inferior to men by many nations. It indicates that we are still at a primitive stage of our evolution.

Communitarianism has grown and is becoming an integral part of how societies are developing. This movement is almost the direct opposite of neoliberalism, as it advocates moral values and collaboration. It goes against the market-led economy and places great value on social cohesion. Social organisation is now an important part of civil society's sustainable development initiatives. It has become a symbol of the bottom-up movements (see the case stories in Chap. 6). Technocracy is an essential factor in the current dominant paradigm due to the technological advancement during the past centuries. Many future scenarios are focused on technocracies, and many have taken the form of dystopias. In the American futurist Ray Kurzweil's classic book "The Singularity is Near" from 2005, robotics become part of everyday life, and biotechnology creates eternal life spans. Earlier the political philosopher Murray Bookchin (1921–2006), wrote several books dealing with technology and ecology issues, such as "Post-Scarcity Anarchism" (1971), in which he outlines the opportunities of a post-scarcity society by creating a new society that is based on abundant technological advancement. The ideas compare with those who believe that sustainability can only be reached through technological and scientific advances. Such a believer is the designer and futurist Jacque Fresco (1916–2017). His Venus Project from 1995 advocated a top-down regulated, systemic approach which was very different from the more conventional bottom-up community and grassroots movements. Jacque Fresco's believed that technology could serve humanity to the degree that all mundane work would be digitalised and carried out by robots leaving people free to pursue the arts.

Many books have been written about technology, artificial intelligence, and augmented reality, such as "Utopia for Realists" (Bregman 2017), "The Network Society" (Castells 1996), "On the Future" (Rees 2018), "Time Travel" (Gleick 2017) as well as the pioneering scientist James Lovelock's latest book "The Novacene: The Coming Age of Hyperintelligence" (2019). The reality is that a larger and larger part of our lives is becoming digitalised. Many of the functions previously carried out by people are now computerized and controlled by algorithms. Meetings are done digitally, and our physical presence is no longer necessary. Governments rely on technological inventions to solve the climate change crisis. Not only do they rely on inventions that are in the pipeline, but also on solutions that haven't been developed yet. This reliance, appears to be another political *blindspot* used either as an avoidance tactic or as a symptom of powerlessness. The fact that the current problems cannot be solved through technology alone is probably very challening for politicians to face.

Utopian Experiments

The above has outlined the idea of utopia predominantly from a theoretical perspective to inspire and invoke change. Utopian thinking has also inspired and provoked experimentation often through the construction of new and ideal communities, especially from the eighteenth century onwards. Many utopian concepts were then no longer founded on religious principles, but became more and more secular, political, and later as mentioned above, technological. Added to that were the emerging environmental and ecological problems of the twentieth century.

There were two schools of thought for embedding utopian ideals in real life. One consisted of advancing technologically and integrating new structures in society. The other saw the emergence of small, voluntary, and idealistic communities, often established as an experiment in a place and time context with relevant infrastructure and system for community existence. These communities were established by groups of people who believed in the same ideals and who wanted to live with a specific set of principles and values whether based on religious beliefs, social or political principles. The size of these intentional communities varied from a few families to a few thousand members. Some were established and disappeared after a few years, some lasted several decades, and some still exist today. Among the most widely known are the Quakers, the Shakers, the Levellers, Diggers and Ranters, the Mennonites and the Mormons. Some of these still exist and are based in the US, although many were founded in Europe. Some were founded on religious principles with specific ways of life, and included very radical changes to the current societal order of that time, and some were politically orientated. Despite being closed communities, each left legacies of different dimensions, for example, the classical English shoe brand, Clarks is a legacy from the Quakers. The Shakers have become renowned for their craftsmanship and simple but beautiful furniture designs.

Many of these communities were very powerful at the time, and some still are today. The Diggers were advocates of joint ownership of land and material wealth. They believed in free education for all and advanced innovative methods for cultivation of the land to increase food production. The Mennonites were also one of the earliest community-based and religious settlements, as they were devout protestants and committed to peace. It indicates that it is possible to create societies that are based on alternative philosophies to the existing and that the contemporary paradigm is not the only framework in which societies can develop and flourish. It also shows how experiments with an infrastructure based on sustainability principles is possible, as these intentional communities designed their communities within their own agreed framework. Therefore, how our societies are structured today is not the only way, or necessarily the best way. We have been led to believe it is so by those who benefit the most from the status quo.

The above reminds me of the famous quote by Machiavelli (1469–1527), who has been called *the father* of modern political science and philosophy. His ideas of politics being unscrupulous, manipulative and filled with deception reverberate with many contemporary thinkers in the environmental field, such as Naomi Klein, whose mistrust of corporations shines through her books. The Machiavellian quote below expresses what many feel today, especially about governments' inability to take the necessary measures to alleviate the crises. In fact, the quote encapsulates the current situation perfectly (Niccolo Machiavelli 1469–1527):

> There is nothing more difficult to take in hand, more perilous to conduct, or more uncertain in its success, than to take the lead in the introduction of a new order of things. For the reformer has enemies in all those who profit by the old order, and only lukewarm defenders in all those who would profit by the new order, this lukewarmness arising partly from fear of their adversaries … and partly from the incredulity of mankind, who do not truly believe in anything new until they have had actual experience of it.

One of the most critical utopian theories to have impacted our societies in recent times that has had a global impact was started by two pioneers, Charles Fourier and Robert Owen. They are regarded as the founders of British and French Socialism. Charles Fourier (1772–1837), a French philosopher, was inspired by his contemporary Welsh colleague, Robert Owen. They undertook experiments based on principles that are now described as socialist. Fourier's city experiments or phalanxes, as he called them, were highly organised settlements aimed to hold diverse classes. They were designed to be as varied and satisfying as possible, based on individual desires and passions framed by values of equality and harmony. Fourier saw his city communities as replacements of the cities in France, spaced evenly throughout the country and with a critical mass of fewer than 2,000 members per unit for easy control and regulation (Miles 2008). Fourier's mentor, Robert Owen (1771–1858), was a wealthy manager and social reformer. The American publisher and author George Lockwood, wrote in his book "New Harmony" in 1905 that the two utopians were "social architects fusing enunciation of social theories with actual experiments". Robert Owen developed his ideas regarding the place for work and educational activities in the same way Fourier did in the phalanxes, believing that the more pleasant the environment, the better people would work, learn, and develop. Owen's intentional communities were established first in England, and then in the US.

The socialist ideals lived on and impacted political theory and practice worldwide. Even though some of the ideals have existed since the time of Plato, it was only in the eighteenth and nineteenth centuries that they formed a specific political movement. Not all experiments to establish utopian settlements succeeded. History does not reveal to what degree the failures had a negative impact on the people and the surroundingsor whether there were long-term adverse consequences from the closures of the communities. Without the experiments, however, nothing new would have been tested. Whether they were failures or successes, we can still learn from the experiences gained.

4.4 Sustainability and Utopia

> It is logical to suggest that sustainability is the utopia of the next era.

This next section explores utopia in a sustainability context seen from both a theoretical and from a practical point of view. If, as mentioned earlier, each period in history has its utopia, then it is logical to suggest that sustainability is the utopia of the next era in human history. Many authors and utopians have imagined and developed theories on utopia that contain elements of sustainability, including specific environmental and ecological aspects. The most widely known works include Thomas More's "Utopia", Bernard F. Skinner's "Walden Two" (2005), Ernest Callenbach's "Ecotopia" (2009), William Morris's "News from Nowhere" (1993), Ebenezer Howard's "Garden Cities of Tomorrow" (1985), Aldous Huxley's "Island" (1962) and J. C. Hallman's "In Utopia" (2010). In the following section, I will discuss some of the more widely known ideas from a sustainability perspective.

Sufficiency, Abundance and Other Models

The Dutch scientist and publicist Marius De Geus (1954–2020) divided environmental utopias into two main groups, the utopias of "sufficiency" and the utopias of "abundance".

Sufficiency Models

These models advocated the simple life, back-to-nature scenarios, and ideal places with respect for nature, including the principle of living within the carrying capacity of the earth. Sustainability ideals are not new and seem to have captured the imagination of utopians for centuries, even though these issues were not in focus as such. Thomas More's "Utopia" from 1516 was one of the first models of an ideal society and included elements of what is now called sustainability. His utopia was a society with a focus on the good life. It was a life where waste was not an option and scarcity was dealt with as a serious matter. Nature was respected and revered, and people had a close relationship with the land. Gardens and agriculture played an essential role in the utopian's life. Clothing was simple and understated. The symbols of richness were for the outcasts. Simple living and fulfilment of basic needs prevailed. Community spirit and community living were part of daily life. Production served to satisfy the basics of life, and the main aim was to have enough free time to engage in intellectual and non-material activities. Drawing a line from More's "Utopia" to what many people desire in their societies today, parallels such as the wish for contentment, fulfilment, and happiness are the most essential components (Garforth 2018). The references to sustainability in terms of reduction of excess and the focus on the non-material are part of the sustainable development goals today.

Another utopia of sufficiency was that of Henry Thoreau's "Walden; Or, Life in the Woods" (1995). His utopia was the back-to-nature model, a utopia of simplicity and naturalness. Thoreau was an advocate of sustainability not promoted through an idea of an ideal society but instead through his idea of an individual's life. It was a life that he experimented with and promoted as a model for the rest of us to follow. Thoreau was highly individualistic, a person who walked his talk and lived what he preached. Relating this to today's society, Henry Thoreau could be called a pioneer and a visionary and would be respected because he lived an authentic life. Simplicity was a key value. His ideas were forerunners for, for example, the slow food movement of current times, which started in Italy in 1982 by the politician Carlo

Petrini, who set out to promote locally sourced foods. He developed the slow-food and simple living movement in Italy, a movement that became world famous.

A third well-known sufficiency visionary was the earlier mentioned British textile designer, novelist, printer, and activist William Morris (1834–1896). He was a utopian with a specific focus on the field of art and design. In his book "News from Nowhere" (1890), Morris described his ideal society, based on harmony, balance and beauty. Included in the equation were freedom and justice. Morris was a libertarian believer in art and beauty as essential ingredients in an ideal society. In this, he differed from the ideas of his contemporaries and was a passionate advocate for art as essential for intellectual and cultural development. He also believed in appropriate technology as a key driver in his land of utopia. Morris's environmental approach was embedded in his belief that nature was important but had to be managed and controlled as a garden landscape. Morris also advocated the simple and natural life and promoted values of production that included quality, durability and aesthetics (De Geus 1999) (see also Sect. 4.1).

In the current field of sustainability, aesthetics, experienced through art and design, have, until the last decade, not figured as an element of importance in the discourse. Art and design have been viewed as superfluous and sometimes directly damaging and as an integral part of the un-sustainable world. The disciplines were regarded as part of the cause of environmental degradation and as metaphors for the wasteful production of luxury goods and consumerism. Opposed to this, the architect William McDonough and chemist Michael Braungart, the founders of Cradle-to-Cradle, include both beauty and design and abundance in their sustainability principles. According to Michael Braungart, beauty is not a wasteful concept or value. It is needed to enhance lives and is an essential tool to ease, as well as further, the transition to sustainability—so true.

Autarky is a name for an entity, such as a nation or state that is economically independent or self-sufficient, comparable to the sufficiency utopias. Countries were mainly autarkies in the past, but due to development, globalisation, and the modern economic framework, very few can be called autarkies today. We live in a globalised world. All countries depend on each other in some form or other. The most current example is the energy crisis brought about by the Russian/Ukrainian war in 2022, where most European countries have been dependent on energy such as oil and gas from Russia. Globalisation has, in this context, a sting in its tail as it is problematic to trade with your enemy. It will be interesting to observe whether there will be a shift towards more autarky nations in the future. Part of the sustainability agenda calls for increased self-sufficiency and localization at the same time as collaboration becomes increasingly essential. The challenge will be to combine localisation with a globalised outlook.

Abundance Models

The second main group of environmental utopias was called the utopias of "abundance", where science and technology were regarded as effective tools for creating abundant lives. Scarcity did not exist in these utopias. Utopias of abundance are found in works by authors such as the American journalist and writer Edward Bellamy, who wrote the book "Looking Backwards" (2000). He envisaged a socialist society, where the citizens took turns to serve industrial advancement to make time for cultural activities. Another author was the English writer H. G. Wells ("A Modern Utopia" 2005), who envisioned a scientific and democratic future where education played a major role and where a world government was established. I talk more about technocracies in Sect. 4.3.

Sustainable City Models

Apart from envisaging new intentional communities many pioneers wanted to address the cities. Cities have caught the imagination of numerous utopians or idealists both past and present, especially architects, planners, and developers. During the eighteenth and nineteenth centuries new visions for cities emerged and became the focus for experiments. This interest grew and today numerous cities have become models for sustainable development. The importance of this is, of course, that more than half the world's population now lives in cities. Cities can be viewed as closed entities, where new infrastructures can be explored, and applications can be systemic.

The British urban planner Ebenezer Howard (1850–1928) was one of the first pioneering city developers. Howard outlined in his book "To-morrow: A Peaceful Path to Real Reform", written in the late nineteenth century, a design for ecological garden cities, where people could live side by side with nature in perfect harmony. Howard was against private ownership of land, which he believed was the root of unsustainable living through the destruction of the natural habitat and the rise of social inequity, causing unhappiness and dissatisfaction. He advocated paying rent instead of owning land and property. The rent would cover the cost of the buildings and maintenance, which today might be regarded as life-long taxation. Ebenezer Howard's ideas on the economy were not unlike the new green economy proposals, which are part of the current discourse on changing the economic system. Howard's infrastructure in terms of supplies and waste flows, included ideas such as creating water reservoirs to create hydro-powered electricity and supply fresh water. Waste was treated and recycled and used on fields as fertilizers. Transport was minimized due to locally produced and manufactured food and goods. Howard based his ideas on creating cities for residents with a relatively high standard of living. Even though he did not promote growth directly, he did believe that a relatively high degree of wealth was necessary for people to live healthy and happy lives (Howard 1902). Some of these ideas can be found in current city planning.

Other contemporary authors and architects were inspired by the city's potential, such as the author Herbert Girardet, who wrote "Creating Sustainable Cities" (1999), Malcolm Miles with "Urban Utopias" in 2008 and "Visions of Architecture and Urbanism" by Robert Klanten and Lukas Feireiss (ed.) in 2011. In this context, the world-famous British architect, Richard Rogers, must be mentioned, who died in 2021. He wrote the book "Cities for A Small Planet" in 1998, in which he described his idea of the sustainable city (quote):

> A Just City, where justice, food shelter, education, health and hope are fairly distributed;
>
> A beautiful City, where art, architecture, and landscape spark the imagination and move the spirit; A Creative City, where open-mindedness and experimentation mobilise the full potential of its human resources and allows the fast response to change;

> An Ecological City, which minimises its ecological impact, where landscape and built form are balanced and where buildings and infrastructures are safe and resource efficient;
>
> A City of Easy Contact and Mobility, where information is exchanged both face-to-face and electronically;
>
> A Compact and Polycentric City, which protects the countryside, focuses and integrates communities within neighbourhoods and maximises proximity;
>
> A Diverse City, where a broad range of overlapping activities create animation, inspiration and foster a vital public life.

Buckminster Fuller (1895–1983) must also be mentioned in this context. He was an American pioneer, architect, and system theorist and dedicated his work to improving the human environment and promoting sustainability. He was famous for his geodesic domes, which can be experienced in the Eden Project in Devon, UK.

It became clear to several city mayors in the 1960s and 1970s that could sustainable development initiatives diminish the adverse effects of industrialisation, the cities were the ideal places to start. Inspirational mayors have been the drivers for change in the cities since the 60s, when the mayor of Curitiba in Brazil, Jamie Lerner, was one of the first to introduce sustainable development initiatives. A current initiative is the C40 network of cities worldwide. The network consists of 97 megacities (cities of more than 10 million inhabitants) and a few smaller model cities, such as Copenhagen. The network represents approx. seven hundred million citizens and 25% of the global economy and is committed to addressing climate change and delivering on the goals of the Paris Agreement. "C40 supports cities to collaborate effectively, share knowledge and drive meaningful, measurable and sustainable action on climate change" the website stated in 2019. At a Summit in 2019, it was announced that as governments were not doing what was required to combat climate change, the network's members (the mayors) agreed to create economic and environmental equality and justice in the megacities through the Global Green New Deal. I applaud the city mayors. Many have shown the kind of leadership, courage and commitment that is so lacking in politicians. One can hope that the effect of a worldwide network of committed mayors will create radical change and inspire governments to follow suit.

Ecotopian Models

Some writers have produced ideas of entire sustainable societies, called ecotopias. Ernest Callenbach's book "Ecotopia" from 1975 belongs to such writings. It is one of the most detailed, as well as recent complete description of a utopian model for a sustainable society. Ecotopia is a country which covers a large part of the US after having split from the rest of the country. It encompasses the states of California, Oregon and Washington. The change of the society from being one belonging to the traditional world outside to creating a new order was transformational and system-based.

The new society is built on an ecological infrastructure and includes radical changes to political, educational and economic aspects to include flow-orientated and production-based systems. Ecotopia is an ecologically stable-state society, where all policies and decision-making are taken, considering the impact they have on the natural habitat and living systems. However, there is much social control and a strong moral sense of what people can and cannot do. Agriculture has been changed, whereby all waste products are recycled, and no chemicals are needed or used in the production of food or other agricultural produce. In terms of the industrial output, consumer goods that are harmful to the environment are no longer made. Second-hand furniture and clothing are the trends of the day, and recycling is part of daily life. There is no pollution of air, water or soil. Energy supplies have been changed from fossil fuels and substituted by a variety of alternative energy systems. Large cities and skyscrapers have been broken down and transformed into smaller units of communities interspersed with gardens, forests, fields and orchards. Infrastructure in terms of transport and mobility provides free transportation in underground systems, electric busses, and magnetic levitation trains. These speed through the landscape at up to 360 miles per hour, rendering travel by plane obsolete. Families of 5–20 people replace the 4-unit family, and one can choose to stay in one family or move to other family units. The working hours are short, and cultural and artistic activities are encouraged, as in William Morris's utopia.

As with previous utopias the focus is on maintaining a balanced society, the natural habitat is respected and preserved, and the human footprint is minimized. The difference between "Ecotopia" and other sustainable utopian models is that Callenbach has taken an existing place and transformed it into a sustainable society. Using an existing infrastructure he has changed and adjusted it to fit a new framework. Callenbach does not denounce technology but uses it to serve society and people rather than having it dominate life. Many elements in Ecotopia are practical. Some we see in our communities, such as recycled clothes and furniture. The other elements seem obvious and relevant, and do not need advanced technology to be implemented.

4.5 Experiments with a Sustainability Bias

Many utopias were constructed from the seventeenth century onwards. A previous section mentioned the forefathers of socialism, Robert Owen and Charles Fourier, who established several communities and settlements based on their ideas of the ideal society. Ebenezer Howard's idea of garden cities was also advanced to a stage where the model could be put into practice. Letchworth, to the north of London, was an attempt to build a garden city. Howard founded the Garden City Association in 1899 to generate enough capital to build the Garden City as a master blueprint and model for a series of satellite cities housing approx. 250,000 people. The Garden City Association still exists, now called the Town and Country Planning Association said to be the oldest environmental charity in the UK.

A variety of experiments took place during the past century in the form of movements, settlements, communities, villages, cities, or parts of cities. As with the sustainable utopian models these later experiments can be categorised into sufficiency and abundance groups (see the previous section). The sufficiency-based experiments were founded on deep ecology principles with a strong back-to-nature element, much like the philosophy of Henry Thoreau. This group of communities encompasses hundreds of relatively small settlements and communities with less than 5,000 residents. They can be found on all continents around the world. Most of them are separate from the mainstream, having embraced radical changes in lifestyle in accordance with their philosophies and principles (Miles 2008). They include grassroots communities such as the Aldo Leopold Wilderness in New Mexico, the first rewilding community in the US, and Twin Oaks in Virginia, an egalitarian community, established in 1967 (Hallman 2010), inspired by BF Skinner's model from his book "Walden Two" (1948). In addition, there are spiritual communities like Findhorn in Scotland, founded in the 1980s and Brahma Kumaris, a world spiritual movement with centres all around the world, established in 1937 (see the case story on Brahma Kumaris, Sect. 6.5), the experimental town of Auroville in India, established in 1968 as well as the Ecovillage Network, established in 1999. In addition, social and political communities such as the kibbutz settlements in Israel can be included in the sufficiency group. They are, of course, all different, have developed their particular way based on their cultures, individual ideals, values, and principles, and all have adopted an alternative lifestyle separate from the mainstream. There are numerous intentional communities scattered around the world where people have decided to go their own way, especially in terms of creating sustainable communities. I will, in the section below, mention a couple of examples of the sufficiency models, where one example is a network of communities, and the other is one man's vision for another type of settlement.

The Ecovillage Network is highly regarded and respected by spiritually and ecology-minded people. The Network consists of approx. ten thousand communities globally and has developed an extensive global education platform for refugees and disadvantaged people. It is a worldwide network of like-minded people who have created sustainable village communities, with the main aim to create "a new culture, and to heal the earth" (Global Ecovillage Network 2019). The movement is based on a philosophy that advocates living on the earth with respect for and in balance with all beings and natural systems. They are a model of how several sustainable utopian experiments have been implemented in the twentieth century through a shared philosophy and vision comparable to the kibbutz movement and the social settlements of Robert Owen. The ecovillages are built from a variety of designs, that have been adapted to the natural environment and climate conditions of each place. The villages are found on all continents, from Mexico and Columbia to UK, Norway, Denmark, Italy, Germany, Australia, India, and South Africa.

A well-known, small, and different settlement from the above is Arcosanti in Arizona, US. It was one of the first recent utopian villages created by an American /Italian architect, Paolo Soleri in 1970. He developed a building programme and a plan for a village of 5,000 inhabitants. The goal for Paolo Soleri was to experiment with architecture and ecology, and he invented a new concept for a building design called arcology, which is a combination of architecture and ecology. It attempted to demonstrate how a town could be built with a small ecological footprint. The principles Soleri advocated were the combination of social activities, minimal resource use, and accessibility in synergy with the natural habitat. The design was organic, detailed and aesthetic, similar to Howard Ebenezer's Garden City. Arcosanti, however, never became the settlement it was intended to be, and the development stranded, mainly due to lack of funding. Today it is a tourist attraction with only a few families living there (Soleri 1984).

Other experiments aimed to establish societies in balance with the carrying capacity of the earth through the utilisation of science and technology—a mixture of the sufficiency and abundance models. Many modern ideas and scenarios belong to this group as the idea of realising sustainability through science and technology became increasingly prevalent. Many emerged from architects', designers', and engineers' drawing boards to create high-tech sustainable environments for town or city residents, whether in a village, part of a city or an entire city. Examples of these are the New Songdo business district in Seoul, South Korea, the abandoned project to create the sustainable city Dongtan in China, and the sustainable city of Masdar in Dubai, UAE. What links these examples is that the main aim of their establishments was to reduce the negative environmental impact by greening the cities and by enhancing the living and working environments for the residents. The changes to infrastructure were carried out with the aid of technology in areas such as transportation, energy, water and waste, the use of bioenergy, solar, and wind-power. Closed loop systems were introduced as clusters where waste products from one source was utilised by another. However, Dongtan, the sustainable city development outside Shanghai in China, was stopped due to planning problems, too short schedule and a lack of funds. Others did not turn out to fulfil the intended goals. The new Songdo district in Seoul, Korea, was intended to be inhabited by families with children. However, this did not materialise instead, the main occupants in the district became businesses. The other famous city project, the Masdar city development in Abu Dhabi, designed by the British architects Foster and Associates, was due to have been finished in 2008 but suffered an economic setback during the recession in 2007–2008, and the plan to complete the project was delayed (Hallman 2010). From a technological point of view, these projects were successful but lacked a human-orientated perspective.

The success of intentional communities tells us that we can change our environment if we wish to. It is possible to join a group of like-minded people and create a self-sufficient, sustainable community based on the values and principles that work for you. We can learn much from looking closer at the above examples, especially the GEN, the Global Ecovillage Network. More and more people in Denmark, for example, are establishing smaller local communities based on sustainability principles. They are not closed communities as most utopian settlements, but are integrated into the mainstream infrastructure. Within the communities, renewable energy forms, organic food-growing, social interaction, and collaboration play significant roles, and residents share decision-making and responsibilities equally. Denmark has the highest proportional occurrence of small intentional communities in the EU, mainly established by groups of friends, colleagues, or neighbours, who want to design and take responsibility for their way of life.

As the sustainability discourse has become more and more mainstream, the ideas and concepts for a future sustainable society, have become equally so. As described in the chapter, numerous attempts were and are made to drive societal, sustainable development initiatives worldwide, and can act as inspiration for the way ahead.

4.6 Perspectives

> Where development theories advocate reform of existing systems, utopian theories advocate the replacement of the systems.

Some researchers claim that the emergence of utopia is grounded in philosophy. The ideas are "philosophies, fantasies, rationalizations, projections, images and opinions in which people pattern their behaviour" (Mumford 2008, p. 8). Others maintain that it is essentially human nature to reflect on and address issues regarding humanity and evolution and that scholars from a diverse spectrum of disciplines have throughout history been engaged with the question of society, its betterment and humanity's place within it. The same can be said of development theories, but where these advocate reform of existing systems, utopian theories advocate the replacement of the systems.

The role and effect of utopian thought have had a critical impact on societies. They are considered to have been conducive to advancement, politically and socially—and in the twentieth century, also environmentally. It is the negative aspects of society and human conditions, that have birthed utopian thinking either as an escape from an undesirable reality or as an attempt to change what was unacceptable or unsatisfactory. Reflecting on and wanting to improve how people live and operate, appears to be an integrated and enduring part of the human condition. There will always be people who will develop utopian ideas or models for alternative places or suggest radical changes to existing societies. It is therefore not surprising that so many attempts to translate the theories into practical solutions have been undertaken, whether these have been successful or have ended in failure.

Sustainability is a worldwide initiative, systemic, universal, interconnected, and addresses all of humanity. No large-scale society has as yet reached a sustainable state, and attempts to embed the ideals in real life have mainly been confined to small-scale undertakings by civil society, individual citizens, communities, villages and municipalities. These practical undertakings are comparable with the utopian experiments where small-scale was a practical necessity for realisation. Utopian ideas have been created by pioneering thinkers and leaders or groups of like-minded individuals. Sustainability, on the other hand, was brought forth as a conscious and joint initiative by the members of the Brundtland Commission in 1987. It happened in a different context and was based on years of research. The concept of sustainability was not communicated or launched as an inspiring narrative to capture people's imagination. It could have been. I would maintain that the lack of imagined positive visions for a balanced and sustainable world has been a barrier to implementation initiatives. Instead, there has been a focus, consciously or unconsciously, on dystopian scenarios in mainstream society. It is as though the interest and hope for a better world have disappeared from human imagination in the pursuit of consumerism, and where hopelessness and apathy have taken over. All the practical examples of alternative communities mentioned in this chapter that have succeeded are based on a shared vision of a new way of life. If we could catch people's imagination and awaken their inner desires for beauty, well-being and living better and happier lives, the transition process would flow more easily.

Some key concluding points can be gathered from the chapter. Sustainability, and sustainable development can, to a large degree, be regarded as utopian constructs mainly because they imply making a radical societal change from the present to a new and better existence. Regardless of how convenient and undisruptive, it would be to avert the environmental catastrophe with a few reforms, everyone today must realise that this is neither feasible nor realistic. Another conclusion is that drawing from the experience of practical utopian experiments, it is evident that a radical vision cannot be realised within a short time period, whether in relation to establishing new intentional communities, new types of city structures, or developing or promoting political movements. Ideas take time to manifest, grow and mature. The goal of achieving the 17 SDGs by 2030, having been launched in 2015, is wishful thinking. My final points include the observation that small-scale experiments have had greater success and continuation than large-scale efforts, mainly due to the ability to control, direct and finance smaller initiatives. As such, they can be used as a testing ground for larger-scale initiatives. Another point is the importance of visionary leadership. It is an essential factor in creating something new. Visionary leaders often have the will and courage to achieve the goals that are necessary for transition processes. And finally, the importance of positive imagination, as described in the above text, is crucial when developing new radical initiatives.

Readings and References

About Utopia (2012) The society for utopian studies. https://utopian-studies.org/. Accessed 14 March 2012

Blincoe KS (2020) Essential sustainability: from utopian theory to practical application. Dissertation, University of Kingston. https://eprints.kingston.ac.uk/id/eprint/51146. Accessed 2020–2022

Bookchin M (1971) Post-scarcity anarchism. Ramparts Press

Bookchin M (1980) Toward an ecological society. Black Rose Books, Montreal, p 280

Bregman R (2017/2014) Utopia for realists and how we get there. Bloomsbury Publishing, New York

Brundtland Report (1987) Our common future. https://sustainabledevelopment.un.org/content/documents/5987our-common-future.pdf. Accessed 10 Jan 2019

Butler-Bowden T (ed) (2012) The republic: the influential classic. Oxford University Press, Oxford

C40 Network (2022) C40 cities. https://www.c40.org/about-c40/. Accessed 20 April 2022

Callenbach E (2009) Ecotopia, 4th edn. Bantam Dell, New York

Carey J (ed) (1999) The Faber book of utopias. Faber and Faber, London

Carson R (1962) Silent spring. Haughton Mifflin, US

Castells M (1996) The rise of the network society. Blackwell Publishers, Cambridge

Claeys G (2011) Searching for utopia: the history of an idea. Thames & Hudson, New York

Claeys G (2018) Dystopia a natural history. Oxford University Press, Oxford UK

Claeys G, Sargent LT (eds) (1999) The utopia reader. New York University Press, New York

Club of Rome (2019) Limits to growth. https://www.clubofrome.org/news/publication-of-the-limits-to-growth/. Accessed 10 Oct 2019
Colborn T, Dumanoski D, Myers JP (1997) Our stolen future, 2nd edn. Abacus, London
Collins Dictionary (2012). Harper Collins, Glasgow
Concise Oxford Dictionary (1982). Oxford University Press, Oxford
Crutzen P, Stoermer EF (2000) The Anthropocene. IGBP's Glob Change Newslett (41)
Dawson J (2006) Ecovillages: new frontiers for sustainability. Green Books, Totnes
De Geus M (1999) Ecological utopias: envisioning the sustainable society. International Books, Utrecht
De Geus M (2002) Ecotopia, sustainability and vision. Sage Publications Ltd., New York
Diamond J (2011) Collapse: how societies choose to fail or succeed, 2nd edn. Viking Penguin, New York
Fishman R (1982) Urban utopias in the twentieth century. MIT Press, Cambridge
Fresco J (1995) The Venus project: the redesign of a culture. https://www.thevenusproject.com. Accessed 10 Nov 2015
Fuller B (2008) Utopia or oblivion: the prospects for humanity. Lars Muller Publishers, Zurich
Gaia Villages (2019) Findhorn. http://www.findhorn.org. Accessed 10 March 2019
Garforth L (2018) Green utopias: environmental hope before and after nature. Polity Press, Cambridge UK
GEN (2022) Global ecovillage network. https://ecovillage.org. Accessed 20 April 2022
Girardet H (1999) Creating sustainable cities. Green Books, Totnes
Gleick J (2017) Time travel: a history. Vintage Books, NY
Goodwin B (ed) (2001) The philosophy of utopia. Routledge, Abingdon, p 3
Goodwin B, Taylor K (2009) The politics of utopia. International Academic Publishers, Bern
Hallman JC (2010) In utopia. St. Martin's Press, New York
Harding S (2006) Animate earth: science, intuition and Gaia. Green Books, Totnes
Harvey M (2019) Utopia in the Anthropocene: a change plan for a sustainable and equitable world. Routledge, Abingdon, p 6
Hillton J (1933) Lost horizon. Macmillan, London
Howard E (1985) Garden cities of tomorrow. Swan Sonnenschein & Co., UK
Howard E (2010) To-morrow, a peaceful path to real reform. Cambridge University Press, NY
Huxley A (1946) Brave new world. Harper & Brothers, New York
Huxley A (1962) Island. Chatto & Windus, London
Jackson H, Svensson K (2014) Ecovillage living. Green Books, Totnes
Joubert K, Dregger L (2015) Ecovillage: 1001 ways to heal the planet. Triarchy Press, Axminster
Kaku M (1998) Visions. Oxford University Press, New York
Klanten R, Feireiss L (eds) (2011) Utopia forever. Gestalten, Berlin
Kropotkin P (1899) Fields, factories and workshops. Houghton, Mifflin & Co., London
Kurzweil R (2005) The singularity is near. Gerald Duckworth & Co., Ltd., London
Levitas R (2013) Utopia as method: the imaginary reconstitution of society. Palgrave Macmillan, Basingstoke
Lipton BH (2004) The biology of belief. CPI Group Ltd., Croydon
Lockwood GB (1905) The new harmony movement. D Appleton & Company, New York
Lovelock J (2019) The novacene: the coming age of hyperintelligence. Allen Lane, UK
Mannheim K (1936). Ideology and utopia: an introduction to the sociology of knowledge. Harcourt, Brace & Co, New York
Miles M (2008) Urban utopias, the built and social architectures of alternative settlements. Routledge, Abingdon
More T (2010) Utopia. Feather Trail Press, USA
Morris W (1890/1993) News from nowhere and other writings. Penguin Books, London
Mumford L (2008/1922) The story of utopias. Forgotten Books, UK
Norberg-Hodge H (2000) Ancient futures: learning from Ladakh. Rider, London
Orwell G (1949) Nineteen eighty-four. Secker and Warburg, London

Pappas N (1995, 2003) Plato and the republic, 2nd edn. Routledge, New York
Rees M (2018) On the future. Princetown University Press, New Jersey
Rogers R (1998) Cities for a small planet. Westview Press, Colorado
Rybczynski W (2010) Makeshift metropolis: ideas about cities. Scribner, New York
Shredroff N (2009) Design is the problem. Rosenfeld Media, New York
Skinner BF (2005/1948) Walden two. Hackett Publishing Co. Inc., Indianapolis/Cambridge
Smith A, Darnell T (eds) (2018) The wealth of nations. 12th Media Services
Society for Utopian Studies (2020). Utopia. https://www.utopian-studies.org. Accessed 20 April 2020
Soleri P (1984) Arcosanti, an urban laboratory. Avant Books and Cosanti Foundation, San Diego
Steel C (2009) Hungry city: how food shapes our lives. Vintage Books, London
Thoreau HD (1995) Walden; or life in the woods. Dover Publications, NY
Utopian Studies (2020). http://www.psupress.org/Journals/jnls_utopian_studies.html. Accessed 4 March 2018
Vonnegut K (1952) Player piano. Charles Scribner's Sons, New York
Wells HG (1905/2005) A modern utopia, 2nd edn. Penguin Group, London

Chapter 5
Implementing Sustainability in the Current Paradigm

Abstract This chapter discusses implementation initiatives and assesses the overall achievements and learning harvested from both top-down and bottom-up global undertakings to implement sustainability since 1987. A great deal of research has been focused on the development of a variety of application methodologies in the form of principles, models, methods, indicators, monitors, and assessment tools. As the world was waking up and starting to understand the seriousness of the emerging problems and challenges, the sustainability discourse became a topic on many corporate, business, and political agendas. Scientific and other academic research into sustainability and new technologies started to flourish. Many organisations, research institutions, NGOs, and pioneering individuals worldwide worked with the concept and developed frameworks and tools that could be used for sustainable development activities. The chapter examines these endeavours, covering some key implementation tools that grew from academic research and from business sectors. Numerous models have seen the light of day since the Rio Summit in 1992, and new models are continually being added to the list. The chapter will document some of the most critical models as well as examples of assessments and standards to give an idea of the broad spectrum of implementation attempts and the complexity that they contain.

5.1 Implementing Sustainability

> The concept of sustainability is fairly straight forward. Achieving sustainability in the real world presents a daunting and complex challenge (Kibert et al. 2012, p. 2).

The chapter addresses critical implementation initiatives from 1987 until today, covering principles, models, processes, and methods, as well as assessments, indexes, and standards. Reflecting on societal transitions in the past, discussed in the last chapter, only a few transformational changes have taken place over short periods (decades rather than centuries). Only revolutions, political coups, wars, and recently the pandemic have happened suddenly. Peaceful societal changes have, in the main, occurred over long time periods and have been of a reformist rather than a radical nature. The era of industrialisation has lasted a couple of centuries, as has the capitalist system. Integrating new political, economic, or or societal beliefs takes time.

K. Blincoe, *Achieving Sustainability: The Ultimate Human Challenge*,
https://doi.org/10.1007/978-3-031-10023-9_5

Smaller intentional communities have been able to move faster, initiating experiments according to new sets of principles and visions.

The experiments in the previous chapter are indicative of this. The larger the entities, the more complex they are, as they are made up of systems and infrastructures that embrace a broad range of diverse societal aspects. To some degree, this explains why so little has changed over the past three and a half decades in terms of development in a sustainability context, despite the many initiatives and activities. Another element causing the lack of progress is that there has been no tried or tested framework or action plan to initiate the required transition, a transition of global magnitude. I have read statements by people saying that we cannot use the experience gained from history when moving forward this time. I disagree with this. Much can be learnt from how humanity has dealt with extreme challenges such as wars and extreme natural disasters. We can also learn from smaller and similar events. We can listen to and learn from those who have already survived the impacts of environmental disasters around the world and who have moved on. We can find out how they survived, what they did, and what worked. Similarly, we can learn about processes, such as the three case stories in Chap. 6, regarding the factors necessary for successful transitions as well as the factors that create barriers. Many experts on climate change have ideas on how to move forward with innovative and new processes. But there is no guarantee of success from them either. However, that should not keep us from trying, as we must. But we must expect and accept that a transition as deep and wide as the one we have embarked on will take time, and the cost will be felt by everyone. To guide us forward, I believe it is essential to look at what we have done so far—in this instance, since 1987. How have actors locally, nationally, and internationally attempted to implement sustainable development initiatives, and what has been achieved?

5.2 Principles, Models and Processes

The Brundtland definition of sustainability and the subsequent Rio Declaration in 1992 intended to guide countries towards sustainable development. Unfortunately, there were no explicit guidelines or tools established by the UN regarding the implementation of the ideas and principles—only non-binding declarations. No how, when, where, or what apart from the recommendations deriving from the summits. The concept and application processes were left open in terms of interpretation and implementation. They were picked up by the few, who at that time were sufficiently aware of the possible perils of the continuing non-sustainable behaviour. After the launch at the Rio Conference in 1992, some sectors in society began working with sustainability, including government sectors, community networks, and individuals, who started to initiate change nationally and locally. To begin with, the pioneers and passionate believersled the drive from the bottom-up. For example, the Samsø initiative (Sect. 6.2) which started in 1997, was headed by local pioneers. Another example was the Transition Movement, initiated by one individual, Rob Hopkins, in

2006 in Totnes, Devon, UK. This movement spread to many towns around the world. I mentioned several pioneers in the chapter on Utopia. Still, there have been numerous innovators around the world, who have spearheaded amazing practical initiatives to further sustainability. They have established organisations, campaigns, events, and local and global movements, and have worked tirelessly for the sustainability cause.

Principles

The first and least complicated exercise after 1987 was to create sets of principles to introduce sustainable development into organisations, businesses, or similar establishments. Multiple principles were created by public and private institutions, the business sector, and civil society. Different actors developed sets of principles around their specific contexts. There were many variations on the theme of what sustainability or sustainable development was supposed to imply. It is clear from the many and varied principles for sustainable development that there was a lack of clarity and information from the start, which has not helped the implementation process. In this context, I want to mention the American researchers Heather M. Farley and Zachary A. Smith, Professor of environmental and natural sciences. They authored a book published in 2014 called: "Sustainability, if it's everything, is it nothing?" The title reflects a critique and sentiment relating to the term sustainability, that is shared by many in the mainstream sectors. Because, as the authors say, the term is overly used. There are so many opinions on the meaning of sustainability that the construct has become diffuse. I have heard this said many times and believe that the reason is due to a general lack of knowledge about the term and what it means. Farley and Smith advocate a new term *Neo-Sustainability* and end their book by defining the phrase, which includes three factors: that there are limits to growth, that these limits are related to the natural systems, and that neo-sustainability is based on systems thinking. I am perfectly happy with the existing term and believe it covers the necessary aspects, but I understand the confusion. A confusion that could easily be alleviated by introducing sustainability in the school curriculum (see Sect. 7.1).

I examined a total of 105 principles in the research for my thesis, created by a wide variety of public and private organisations between 1991 and 2015, such as WWF, The Worldwatch Institute, NRTEE (National Round Table on the Environment and Economy, UK), International NGO Forum (UN), APEC Ministers responsible for the Environment (Asia), Centre for Policy Alternatives (US), United Kingdom Secretaries of State for the Environment (UK) as well as several research centres and private companies. The principles imply that the task of implementation not only involves a top-down approach through legislation but involves everyone. I have only chosen to record a couple of sets of principles as they were all very similar. They followed the same outline with a list of approximately ten points relating to the field of work of the originators, and they were nearly all general rather than specific. Principles have been used as guidelines rather than as specific goals, and as such their prime merit seems to be that they have spread awareness, if nothing else.

The Department of Agriculture of the US government launched a set of Leadership for Sustainability Principles for their employees in 2010. These included:

- Leading by example even if this requires a new direction.
- Using the appropriate and relevant indicators and measurements of progress available.
- Looking at consumption in an interconnected and integrated way.
- Promoting innovation and resilience.
- Looking for effective *leverage points* to affect change.
- Using biomimicry as a framework for sustainability.
- Working within existing contexts, co-operating, and collaborating.
- Implementing sustainability as a lifelong process.
- Acknowledging that both failures and successes are part of the process. Developing the ability to embrace change and chaos.
  (Worldwatch Institute. The website was accessed in 2015/2016 before the institute closed in 2017).

The principles stressed the individual leader's approach to the sustainability process, advocating the development of specific personal attitudes in the internal working environment and in external relations. It was a list of suggestions for the individual taking a leadership role and for groups of people in business or organisations. The set included elements of value and attitude approaches and was an illustration of a human-orientated call for action, unlike the many vague and broad examples I examined. This set of principles was unusual due to its individual approach. Most principles were top-down-orientated, generalist, and open to interpretation, and hence more difficult to act on.

The principles from One Planet Living originated in 2008 and included a set of ten principles for sustainability, and was a global initiative developed in a partnership between Bioregional, an enterprise set up by the Bioregional Development Group, and WWF (World Wildlife Fund).

The principles were the following:

- Zero carbon emissions.
- Zero waste streams.
- Sustainable transportation.
- Local and sustainable materials usage.
- Local and sustainable food production.
- Sustainable water management.
- Land-use and wildlife protection.
- Culture and community activities.
- Equity and local economy development.
- Health and happiness initiatives.

This set was more environmental and community-focused, as WWF and Bioregional were NGOs, with the environment and nature conservation at the core of their work. The list directed the focus to specific areas for action. It was broad in its reach but did pinpoint specific areas of the sustainability agenda to act on. The complexity of sustainability is clearly illustrated in these ten principles as they cover areas such as transportation, materials, food, water, wildlife, community, culture, health, and happiness (Bioregional 2019). William McDonough and Michael Braungart (Sect. 4.4) developed the Hannover Principles in 1992. The partners founded a company called Cradle-to-Cradle (Sect. 5.2), famous for its ideas on recycling and up-cycling models. The principles were a comprehensive set of guidelines relating to the design of buildings and their environment. The principles included the following:

- Insist on the right of humanity and nature to co-exist in a healthy, supportive, diverse, and sustainable condition.
- Recognize interdependence.
- Respect relationships between spirit and matter.
- Accept responsibility for the consequences of design decisions upon human well-being, the viability of natural systems, and their right to co-exist.
- Create safe objects of long-term value.
- Eliminate the concept of waste.
- Rely on natural energy flows.
- Understand the limitations of design.
- Seek constant improvement by the sharing of knowledge (Blewitt 2008).

The Hannover Principles focused on design and architecture. They also included "spirit and matter", which at that time was unusual. The principles can, as such, be regarded as having a more holistic approach than many of the other principles of sustainable development. Scientists, businesspeople, and academics have started in recent years to recognize the importance of the issues of spirituality in the context of sustainability. Anita Roddick, the CEO and founder of the Bodyshop (1942–2007); Janine Benyus, innovator, author, and the founder of biomimicry; Gregg Braden, author, scientist, and pioneer in healing, are among those who included the spiritual dimension in their work with sustainability. Spirituality is an integrated part of the holistic view of life. When included in the principles, the path towards sustainability becomes more focused on the individual or local way of life rather than the system or infrastructure. As such, I regard it as an essential inclusion. The observations from the case stories in Chap. 7 also indicate that to achieve positive results at a local level, spirituality and matters of the soul are important for the individual and for the communities.

A clear pattern emerged from the examination of the many principles. Most consisted of around ten elements and were short and easily understood. Although described with different words and emphasis, they covered similar core aspects, namely environment, waste, energy, and social issues. Other elements referred to particular work contexts. The sets of principles were variations on a theme. They all appeared to link to the definition of sustainability, whether about development in general, policy development, recommendations for countries or states, or communities and cities. In addition, whether promoting *living* sustainably or *developing* sustainably, they were similar and addressed nature, resources, waste, energy, reduction of CO_2 emissions, and building resilience. Universal and top-down principles were also developed and created by multinational organisations and research centres. They had a wider reach and more non-specific content.

The effect of the work with sustainability principles has not been measured. There has been no evidence of whether the principles examined were integrated successfully or not or whether activities of the host body changed as a result. In the context of principles, it is worth noting that only one of the case studies in Chap. 6, Brahma Kumaris, had a defined set of principles. The other two cases did not utilise principles in their work. Instead, they had a vision, mission and goals combined with a set of values to help guide the processes.

Models and Processes

As with principles, many models have been generated over the past three and a half decades. Most have been research-based, and some have been practical constructs based on experiments. Some were aimed at governments and businesses, and others at organisations as well as civil society. New models, such as the 17 SDGs, are continually being added to the list as the learning and experience increase. Among the most well-known models are Corporate Social Responsibility (CSR), The Triple Bottom Line (TBL), The Natural Step (TNS), The Ecological Footprint, Cradle to Cradle (C2C), Circular Economy and Permaculture, and recently the 17 SDGs (see Sect. 3.5). Most models have been anchored in specific areas related to the expertise, interest, and field of work of the creators, just as the principles. The models addressed below are regarded as some of the classic approaches to implementing sustainable development. Some address the latest emerging scenarios and give an idea of the wide spectrum of concepts that have been developed, as well as the complexity that each carries within it. The models are therefore an indication of the multi-layered dimensions that sustainability represents. The feedback from the application of the models defines the type of issues that have emerged from their application, both positive and negative.

CSR

Corporate Social Responsibility became a mainstream approach towards the end of the 1960s and early 1970s when social and environmental issues became more widely known. Its popularity grew inspired by the work of R. Edward Freeman, an American philosopher and professor of business administration. He wrote the book "Strategic Management: A Stakeholder Approach" in 2010 and became famous for his views on what and who counts in the corporation. CSR was adopted as a strategic tool to inspire and introduce the corporate world to social responsibility in the context of the development and globalisation crises. CSR outlined methods on how corporations could embrace sustainability, the areas to focus on, and the positive outcomes in terms of the business advantages that could be expected. The nature and essence of corporations are about making and securing profit and dividends to their shareholders and attracting investment. CSR goes beyond that to also include both environmental and social aspects. The CSR model deals with corporate stakeholders instead of shareholders. Stakeholders include the internal and external bodies that corporations have dealings with, for example, employees/staff, shareholders, customers, suppliers, retailers and local organisations. The aims of CSR are for the corporation to be inclusive and take responsibility for creating a participatory and respectful way of dealing with their stakeholders. CSR advocates corporate environmental responsibility through paying attention to the environmental footprint, transportation, supplies, and end-of-life of its products, recycling, waste handling and, risk assessment (Henriques and Richardson (eds) 2004). Many corporations and businesses adopted CSR at the end of the twentieth century, and most company websites would include a page that described the company's CSR activities. On the negative side, CSR activities have often become a form of *greenwashing*, which meant that CSR was an easy and non-committed way for a corporation to be seen to be engaging in sustainable development when very little in this regard was being done. Greenwashing consisted of, in simple terms, having a section on the company website outlining the environmental policy and activities that were rarely implemented. Unfortunately, this is still symptomatic of some companies' engagement with the sustainability agenda, especially with the new approach, called ESG, that has taken over from CSR, described below.

The new corporate *black* that has taken the place of CSR is ESG (Environmental, Social and Corporate Governance). ESG differs from CSR in its approach to intangible behaviour internally and externally. Where CSR deals with a corporation's relationship with its stakeholders, ESG constitutes a kind of evaluation or barometer of the degree of social and environmental factors that exists within the company. It is becoming increasingly widespread in the business and financial industries globally. I consider ESG an extension of CSR, as both deal with the ethical and responsible behaviour of a company or corporation. The problem with the latter is that there are no globally agreed standards to measure the results as yet, and that the lack of agreed disclosure rules leaves the system open to being used for *green-washing*.

Triple Bottom Line

One of the most well-known and classical models is Triple Bottom Line (TBL). It has been the most widely used of all models, although Circular Economy has taken over as the model, most businesses currently engage with. The TBL model was mainly developed to deal with corporations and was adopted to drive the sustainability agenda mainstream. John Elkington, a British author and expert on corporate responsibility and sustainable development coined the term the Triple Bottom Line (TBL) in 1994 inspired by the Rio Conference in 1992. TBL has also been referred to as PPP: People, Planet, Profit or EEE: Ethics, Ecology, Economy. The Triple Bottom Line (TBL) model is a type of evaluation model where the profit and loss, people, and planet accounts are in balance. Seven drivers are employed and addressed that deal with markets, values, transparency, time, life cycle technology, partnerships, and corporate governance. The TBL approach became, over the years, a popular model for implementing sustainable development initiatives.

John Elkington described in the book "Cannibals with Forks", published in 1999, what he meant by TBL, and talked about "the sustainable capitalism transition". He believed that businesses would be in the driving seat pushing the sustainable development agenda ahead of governments and NGOs. John Elkington was correct in his assumption concerning governments. As soon as the business sector began to see new market opportunities in the field of sustainable development, they started to become advocates of innovation in the green transition process. The business sector has subsequently played a significant role in the process of creating the 17 SDGs, which were developed in a UN context by a diverse, international working group. As industry began to see the benefits, it was keen to be seen to take a leadership role in driving the development goals. Many corporations are now actively implementing sustainability in their companies, and at COP 26 in 2021 in Glasgow, the business sector played a key role, participating in most of the high-level panel debates and working to produce the financing required. World Economic Forum estimates that it will cost 4–5 trillion $ per year until 2030 to finance the transition to net-zero emissions (Sect. 6.8). That is three times more than previously estimated, and it presents a challenge to the global financial sectors. The business and financial sectors are often ahead of other sectors. There is a logical reason for this, as competition and being ahead are integral to the capitalist ideology and is an integral part of competition and doing business. The green transition offers multiple new opportunities, and the most entrepreneurial actors are already out there creating new kinds of businesses to meet the new demands. Novozymes (Sect. 6.8) applied the TBL model when embarking on their sustainability transition at the beginning of the process.

The Natural Step

The Swedish scientist and cancer specialist Karl-Henrik Robèrt created the framework for a strategic and systemic model called The Natural Step in the late 1980s. The concept stemmed from his study of cancer patients and cancer cells. How and what causes cancer and how the cells grow and develop including the factors that affect and influence their growth. Robèrt's document on The Natural Step was considered so significant that it was broadcast in 1989 on TV and sent to households and schools in Sweden with permission from the Swedish King. Since then, the concept has spread to the rest of the world and was adopted by leaders, corporations, local communities, schools, and colleges as a useful model and methodology to work with sustainable development. The Natural Step advocated four basic system conditions as well as step-by-step approach to implementation of sustainability. The conditions included the following:

- Eliminate the build-up of the substances which are extracted from the earth's resources.
- Stop adding to the toxic substances, for example, dioxins which pollute the earth.
- Eliminate the physical degradation of the natural habitat.
- Eliminate the actions that undermine people's capacity to meet their basic needs in terms of living and working conditions.

In addition, TNS promoted a learning process, a step-by-step methodology to facilitate and help organisations plan for sustainability. The TNS stressed that we are an integral part of nature, not separate from it, and that we cannot exist without clean air, clean water, fertile soil, and diverse natural habitat. When TNS was first created, it gained attention from many levels of society. Especially Scandinavian corporations and companies worked with the model and adopted the methods. However, not much current information exists on the model, and it seems to have receded behind other more recent initiatives (Robèrt 2002).

C2C Model: Waste = Food

The Cradle-to-Cradle (C2C) model is a concept that differs from those mentioned above as it focuses on waste and resource cycles. C2C advocates that waste is a resource that has value and must either be upcycled and made into new resources or broken down into biodegradable waste. All production and consumption should be designed with the two recycling streams in mind. The C2C model was developed by William McDonough and Michael Braungart. They co-authored the Hannover Principles that laid the foundation for their work, the Cradle-to-Cradle methodology. The C2C values include abundance, creativity, prosperity, and change. The key message is that waste equals food, meaning that any kind of waste is a resource to be reused in production processes on par with virgin resources.

The two main elements of the C2C model consist of the biological as well as the technical/industrial/mineral metabolisms or cycles. As a metaphor for C2C, the authors use the tree, where all waste, fallen fruit, and leaves are biodegradable. No harmful residues are left in the soil that can damage the tree, its ecosystems, or surroundings. Instead, the fallen fruit and leaves nourish and enrich the soil. The industrial metabolism or cycle, on the other hand, includes up-cycling as opposed to down-cycling of waste materials. When waste materials such as paper, glass, plastics, or metals have been recycled, they get broken down and become less valuable or useful in new production processes and have a relatively short lifespan. C2C advocates the use of design, materials, production, and recycling processes that enhance the resources, so they can more easily enter into new production cycles (McDonough and Braungart 2002). The C2C model is a focused and specific method to embrace resources and processes. It has been adopted by governments, regions, and cities, and an increasing number of worldwide corporations such as Nike, Herman Miller, and Interface are some of the most well-known who have engaged with C2C.

No model is perfect. One challenge of C2C lies in the complexity of the breakdown, analysis, and disassembly process of waste in the industrial production processes. The difficulty is to develop non-toxic and new materials that can enter the production stream, which can be recycled and reused in new production lines at the end of life or be placed in landfill sites as biodegradable waste. Another problem with the model is that it does not advocate the concept of a reduction in consumerism. This contradicts the classic sustainability measures, which support a drive to reduce production, consumption, resource use, and the generation of waste. The business industry has favoured C2C due to its industrial orientation. Understandably. However, we also need to re-learn how to consume and live with less.

I met Michael Braungart a few times in Denmark and the UK, where he lectured on C2C. The particular inspiration he brought was that he (and his partner, William McDonough) advocate the importance of beauty and aesthetics as essential parameters in C2C design thinking. No other model for sustainable development does that, not even the recent 17 SDGs, which are devoid of any form of cultural content. A crucial mistake, I believe. Everyone understands beauty. It enhances our lives and increases well-being (see Sect. 5.2). Culture is the glue that makes a society coherent and humane. It is the human dimension that is essential for a good life. That includes the senses, the tactile, the aesthetics and the beauty. As such, it should be an integral part of all models for sustainability.

The Circular Economy Model

A currently popular business model, comparable to Cradle-to-Cradle, started to be adopted by businesses, regions, cities, and even countries 10–20 years ago. It is the Circular Economy model. It could be called an extension or development of the C2C concept combined with what a few decades ago were called competence clusters. One of the earliest and more well-known examples of a competence cluster was set up by a group of companies in the city of Kalundborg in Denmark. Six large companies created a cluster, using each other's waste products and feeding them into their own production streams. Waste heat from one company was used in the production processes of another company, whose waste water was cleaned and utilised by a third company, the waste product of which was reused in the production of road surfacing materials by a fourth company, thereby creating a closed circular process or loop. These closed-loop systems (called industrial symbiosis) became one of the hallmarks of the Circular Economy model. Other key ingredients are, as with C2C, the two waste cycles, the mineral-based and the biological cycles. The model requires that production and the built environment are designed so that waste can be reused in new production cycles without losing value and quality (up-cycling). Or be designed with built-in biodegradability in mind, similar to the C2C model (Ellen MacArthur Foundation 2022). From a sustainability perspective, the model lacks the social factor entirely. It is 100% technology and resource focused. The model, as with C2C, does not suggest a change in the consumption patterns that we know are needed to maintain resource and planetary equilibrium. Neither does the model address other ways of doing business, for example, renting, sharing, repairing, lending, exchanging, or bartering. Relating the latter two models to the Brundtland definition, C2C and Circular Economy only address part of the requirement for sustainability. They may be good at managing resources in a lifecycle process and be good for business but do not promote the social aspects of sustainability.

Most of the models above have been developed by and for the business sector. The business sector must be part of the solution, meaning that a fundamental shift in the essence of doing business is required, whether industrial, agricultural, or cultural. Implementing the models described above is still ongoing. Some older models have been replaced. Some are still in the process of development, and new models have been introduced, particularly the Circular Economy model and the 17 Sustainable Development Goals. Corporate business is engaging in driving sustainable development as it has become a good business and aims to secure markets in the future as well as attract investment. Kate Raworth, a visionary and inspiring economist and researcher (Sect. 3.6), states in her book "Doughnut Economics: Seven Ways to Think Like a 21st-Century Economist" (2017) that the key challenge for governments and the business sector is to deal with their continued growth mantra. Raworth believes that we should become "agnostic about growth, creating economies that enable us to thrive, whether or not they are growing". At the beginning of the book, she states: "The most powerful tool in economics is not money, nor even algebra. It is a pencil. Because with a pencil you can redraw the world". Kate ends the book by encouraging us to pick up the pencil and start to draw. I agree. We need to begin to draw and redraw our societies and our lifestyles, and use our imagination to create the society we wish to be part of, one that sustains us and all other living species.

Whole Earth Models: Permaculture

The word permaculture first appeared in the mid-seventies, created by two Australians, Bill Mollison, a scientist and biologist and David Holmgren, an environmental designer. Permaculture is essentially about ethics, based on a set of universal design principles that cover whole-systems and web-of-life thinking. It mimics the way nature creates patterns and interrelationships, building diversity, stability, and resilience in doing so. It integrates several disciplines, including biology, ecology, agriculture, gardening, architecture, technology and community building, and is holistic. Experts on the topic claim that the model is applicable to any system, such as architecture, design, technology, economics, education, and built environments. It has, however, mainly been applied to agriculture, food production, and the natural environment. Permaculture has specific tools that people can utilise. The system is nature-based and can be compared to biomimicry, a design process embedded in and copying nature (Holmgren 2011).

Critics of permaculture include those who maintain that permaculture is a theory rather than a practically applicable model for transition. It is said to lack controlled trials and measured experiments to test what is claimed to be the low input/high output yields as well as the resulting benefits and practical applications in wider society. Some critics see the model as fluffy and idealistic rather than realistic and applicable. Nevertheless, permaculture and forest gardening—a derivative of permaculture—are becoming mainstream initiatives. The models are for everyone to engage with. Despite the critics, many recorded results exist, such as private households, cities areas, rooftops, and a variety of other spaces. They are being cultivated according to the principles of permaculture and forest gardening in the drive to green the cities with excellent results. Permaculture is often part of sustainability initiatives and bottom-up processes led by local people, communities, or councils and can help to generate small and locally initiated food production systems. These initiatives promote social benefits in terms of community building, volunteering, and local enterprise, promoting ethical and responsible values as such, permaculture appears to be closer to the essence of sustainability than some of the other implementation models.

The Forest Garden

Forest gardening is closely linked to permaculture but is generally practised on a smaller scale than permaculture, which is a design system for land use and resilient lifestyles. Forest gardening is an ancient land use method specifically in connection with forestry. It is a system where planting is done in vertical layers imitating young woodland, usually involving edible species such as trees, shrubs, bushes, and plants, producing a wide variety of herbs, berries, fruits, nuts, etc. It is a brilliant form of gardening, as you can grow a large variety of foods in small areas, not only creating a beautiful kind of larder but also creating a haven that encourages healthy ecosystems for birds and insects. I first came across forest gardening, when working at Schumacher College at Dartington in Devon. Martin Crawford, the British forest gardening expert, had developed a small plot next to the College. The visits to Crawford's garden were magical. It was a natural paradise with a great variety of trees, bushes, flowers, plants, and ground cover in all sizes and shapes, all mixed according to their specific mutually beneficial interaction. It led me to initiate creating a forest garden at Schumacher College in 2006/7, developed and run by one of the holistic science students with the help of residents at the College. One lunchtime the following summer, we had a salad made up of numerous different leaves, flowers, and herbs, both beautiful and healthy. Forest gardening is a model and activity, that anyone with access to some small piece of land or garden can learn and gain great joy from. Once established, the forest garden requires little maintenance in terms of weeding and pest control. Apart from being productive in terms of food produce, the garden can become a meditative, restoring space (Crawford 2010).

Rewilding

Rewilding is another method whereby we can restore lost biodiversity systems that are crucial for the survival of humanity and well-being of the natural habitat. It works on a large scale. It is not a model or method to achieve global sustainability directly, but it is an essential sustainability factor that can help repair and replenish nature. Rewilding is more than conservation. It is a proactive system that can restore nature's ecosystem processes through specific interventions such as the introduction or reintroduction of functional species and encouragement of entropy. Rewilding has been gaining more and more interest and focus as the results of the initiatives are spreading and are becoming a movement for positive change, that people can relate to and understand. As part of the conservation movement, it is inventive, positive, and ambitious. Rewilding reintroduces mega-herbivores to the land, such as elephants and rhinoceros in the savanna areas. In Europe, it is Tauros bulls (or aurochs), bison, and wild horses that are reintroduced, as they play a vital role in developing and maintaining biodiversity. The organisation Rewilding Europe embarked in 2013 on a 30-year programme with various partners to recreate a breed similar to the extinct aurochs for introduction into specific areas in Europe. Rewilded land has the potential to store carbon in the soils and recreate ecosystems through the re-establishment of lost flora and fauna (Jepson and Blythe 2020). What kindled my interest was the positiveness and excitement from the people involved with rewilding—a positiveness that is often absent in most discussions on sustainability.

A Universal Model: The IPAT Equation

One of the initial and key attempts to express the human impact on the planet was developed in the 1970s and was called the IPAT equation or formula. The equation was created by Barry Commoner, Paul R. Ehrlich, and John Holdren in 1971. The equation explained the human impact on the environment in terms of three components: population numbers, levels of consumption, affluence, and impact per unit of resource use, which was termed technology. As one component changes so, do the other two. It is, therefore, possible to create an intervention relating to one of the components to achieve a desired result in the others. The equation is expressed as follows:

$$\mathrm{I} = \mathrm{P} \times \mathrm{A} \times \mathrm{T} : \mathrm{I} = \text{ Environmental impact; } \mathrm{P} = \text{Population;}$$
$$\mathrm{A} = \text{Affluence; } \mathrm{T} = \text{Technology.}$$

To reduce environmental impact, the elements that make up the equation need to be adjusted. The IPAT model is, in its essence, a theoretical methodology. It is still quoted today and is used to demonstrate the interconnected aspects of the sustainability factors. Population increase is considered one of the most important factors to be addressed to achieve sustainability, due to food, fresh water, and resource needs (Rosling et al. 2018). It is estimated that the global population will reach 11 billion by the end of the century if the trajectory continues along the current lines. Statistics from the Millennium Goals tell us that an increase in affluence across the globe means greater use of resources. With a population increase and greater affluence, the impact per unit of resource use (T) must decrease dramatically to reduce (I) the impact. In this example, only technology can meet the challenge and lever a balance, as an increase in population and a rise in affluence, impacts the environment adversely.

Relating the equation to the current estimate of an additional 3 billion people receiving middle-class incomes by 2050, it is estimated that CO_2 emissions will rise by 50% if nothing is done to curb emissions in the short term. If governments do not act on the Paris Agreement immediately, the temperature increase can reach 3–5 °C, as previously mentioned. The likelihood of immediate action by governments to step in to limit CO_2 emissions with sufficient green legislation measures is doubtful despite the NDCs (National Determined Contributions) and despite the dire third part of the IPCC 6th Assessment Report which was published recently (Sect. 8.4). The estimated population growth reaching 11 billion by the end of twenty-first century will add a further burden on resource needs. The question is then whether technology can alleviate this burden sufficiently for the impact to be controlled in time to curb global warming. In my opinion, it cannot be done in time, and we must, therefore, learn to adapt to an extreme climate and develop resilience.

5.3 Indicators, Monitors, and Assessment Tools

A critical phase of the implementation process is ensuring that the applied models work according to plan and are generating the desired results. To help further this third level of implementation, several tools, including indicators, standards, certifications, and lately more advanced assessment tools have been generated since 1987. As is indicated by the variety of models and methods mentioned in the previous sections, the task of implementing sustainability is difficult, complex, and requires systemic change as well as innovation at all levels of operations. Research centres, both public and private, have carried out extensive work to find appropriate and operational ways to measure progress and monitor results, and more relevant and practical tools are being produced ongoingly.

The UN, following the Rio Summit in 1992 and as part of Agenda 21, developed the first set of sustainability indicators. Proposals for implementing sustainable development were put forward, and it was logical that these should be followed by a set of indicators to help measure the implementation process and monitor progress. A brief look at some of the indicators explains the underlying difficulty of this field. Many researchers were faced with trying to identify effective and practical ways and methods with which sustainability could be implemented and monitored (Bell and Morse 2008). A few of these are recorded below.

The Bellagio Principles

The Bellagio Principles were one of the earlier attempts at addressing indicators and measuring sustainable development. They were created in Bellagio in Italy by a group of international researchers and experts in 1996. The group proposed ten principles. Four of the ten principles dealt with monitors and indicators on a broad scale, addressing holism, time, and space. They advocated a limited number of indicators and recommended that these had wide accessibility. Communication and participation are also figured on the list. It was an attempt to provide a bridge between the theory and practice of sustainability and thus offers an overview of what was necessary rather than focusing on specific issues. The principles were intended for community groups, NGOs, corporations, governments, and international institutions to initiate and improve assessment activities (Jesinghaus 2014).

The Ecological Footprint Analysis

This analysis was published in the mid-90s by the founders of The Ecological Footprint, Wackernagel and Rees, and was a measure of the amount of (land-based) natural capital that was required to supply a region, a population, or an economy including the resources needed and waste produced. It was developed to measure human impact on the ecosystem. An Ecological Footprint calculator was created and is used by mainstream media for people to try out and measure their impacts. The most well-known action was calculating the Earth Overshoot Day, the day when the global resource output balances with the Earth's capacity to contain it without damaging the natural resource base. The latest calculation is that humans use the equivalent of 1.75 of the Earth's resources p.a. In 2021, the Overshoot Day was July 29th. On this day, humanity had spent its natural resource budget for the year, meaning that for the rest of the year, the resource base would be eroded. Given that society depends on the natural habitat for its survival, the overuse of the natural capital is undermining the future existence of humanity. A parallel can be drawn to Jared Diamond's book "Collapse" which I mentioned in Chap. 2, where he documents how a society's decline and fall can be triggered by a scarcity of resources.

The Environmental Space (TES)

TES was developed by Friends of the Earth in the Netherlands in 1992 as part of the Sustainable Netherland's Action Plan. TES is a methodology that measures the maximum space, a person, region, or country can use for living and working activities, that falls within the carrying capacity of the earth. It is comparable to The Ecological Footprint mentioned above, but its focus is on the amount of space available rather than on the impact caused.

The Life Cycle Assessment (LCA)

This assessment method has been available for many years. It is a methodology extensively used by the production, building, and design industries. LCA is considered to be one of the most frequently adopted assessment tools in the sustainability development process. Early on, LCA's were fairly simple tools generated to measure the sustainability of a product mainly in terms of resource and energy use. The LCA method now covers all aspects of a product's or a service's lifespan. It can give a complete picture of the whole cycle from raw materials, production, and manufacture to transportation, distribution, sales, use, waste, and final disposal. LCA is also called the cradle-to-grave process. One of the cases researched in this thesis, Novozymes (Sect. 6.8) utilised LCA throughout the company's line of approx. seven hundred products to determine how resources were being employed and waste generated. It was one of the critical tools that helped direct the company towards engaging in sustainable practices. Michael Hauschild, Professor in Quantitative Sustainability Assessment at Technical University of Denmark and an expert in Life Cycle Assessment has said:

> We cannot improve what we cannot measure. That is the simple explanation as to why research and method development in Life Cycle Assessment is so incredibly important for the future of the environment and the development of more sustainable industry, transport, energy, etc.

Criticism of the method is that it does not cover social impacts. Another critical aspect is that several LCA methods exist, not just one universal system. Hence, the quality of the specific LCA method applied will reflect the quality of the results generated. The LCA is a science-based methodology like many other methods and has been criticised for being too complex, challenging, and costly for many organisations and companies to benefit from (Kibert et al. 2012).

Sustainability Assessment (SA)

SA is a relatively recent appraisal methodology that encompasses the Triple Bottom Line model of economy, social as well as environmental issues. Sustainability assessment involves the application of several methods, models and indicators and is one of the most complex tools currently available that moves beyond the more specific and technical tools. "Methods to quantify uncertainty are key ingredients of the assessment framework" (Sala et al. 2015).

5.4 Standards and Certifications

The primary purpose of standards and certifications is to ensure that what is produced and how it is produced are *fit for purpose*. ISO standards are essentially created for industries to optimize their business activities ranging from safety to quality, minimize waste and increase productivity and profitability. Many environmental standards in terms of production, strategy and management are generated to ensure that a company's activities in this field follow specific requirements. Businesses are becoming increasingly keen on promoting products and services that have been certified, as it adds a qualitative dimension to their products or services and enhances marketing impacts.

Certification

The certification applies to a wide field of operations, for example, within the sectors of building, business, agriculture, tourism, cities, organisations, and companies. The ISO (the International Organisation for Standardization) is one of the most important standardization organisations. ISO standards are utilised in LCA assessment processes as well as in Global Reporting, GHG protocol, Life Cycle Accounting, and Reporting Standards. The ISO is a large organisation that develops standards for almost every commercialised activity. There are currently approximately 19,500 standards available, but only a few are related to sustainable development. This is about to change, however, as the pressure is on to reach zero emissions by 2050, and new ISO standards for climate adaptation have been launched in the EU countries (ISO 2019).

Labeling

As standards, labeling represents an entire industry. Currently, approximately 500 labels exist in the European market that are related to ecology and sustainability both in terms of material sourcing, production, distribution, social, and ethical issues. The value lies in being able to quickly access information both in the consumer industry, the service industry, as well as in supply chains. One of the barriers is the fact that a company must earn its stamp of approval fulfilling a variety of criteria to gain the use of a specific label. For the consumer, it is a difficult field to navigate in and know exactly what each label stands for. However, labeling of products and services has eased the choices regarding purchases and helps the customer to be reassured that the items bought, such as food products, are responsibly produced (OECD 2020).

5.5 Social Indicators and Indexes

Several social and well-being orientated indicators and indexes have emerged, mainly during the last decade, as human well-being has become an increasingly acknowledged and essential factor in sustainable development. Governments use the GPI

(Genuine Progress Indicator) to examine a country's progress towards sustainable development. Additionally, there is the HDI (Human Development Index), developed by the UN in 1990 to monitor progress on human development. The HPI (Happy Planet Index), was formed by the New Economics Foundation in the UK and is a parallel to GNH (Gross National Happiness) index, based on which the country of Bhutan, for example, measures its progress and development targets. The index is the opposite of the GNP (Gross National Product) which has been the adopted measure of a country's economic progress from 1944 onwards. A further measurement such as the GDI (Gender-related Development Index) from 1995 has offered an alternative perspective but is often criticised for leaving out important components such as ecological considerations, or for being socially or culturally biased (Sharpe 1998; Caradonna 2016). The final index to be mentioned is the Social Progress Index, which was developed in 2010 and launched in 2014. The Index defines social progress as:

> The capacity of a society to meet the basic human needs of its citizens, establish the building blocks that allow citizens and communities to enhance and sustain the quality of their lives, and create the conditions for all individuals to reach their full potential.

The SPI was relaunched in 2015 with some updated content. The aim is to play a major role in translating the Sustainable Development Goals. It is claimed by the initiators that the focus on human well-being will bring about the paradigm shift needed, as opposed to the focus on economic growth. The claim is supported by SPI statistics (2020), showing that economic growth and the degree of human well-being are not interdependent, nor are they proportional. Some countries with medium GDP's have higher rates of human well-being than countries with high economic growth rates. The SPI does not include environmental aspects and can, therefore, not be considered a sustainability index, although social and well-being issues are at the heart of sustainability. In this context, a new partnership must be mentioned that includes Finland, New Zealand, Scotland, Wales, and Iceland called the Wellbeing Economy Governments' Partnership (WEGo), which was set up in 2018. This partnership has several objectives, including increasing the health and well-being of their populations, plus several environmentally orientated aims. Based on this the New Zealand government has, since 2019, used a Living Standards Framework that it developed to track its progress (Well-being Economy Alliance 2022).

5.6 Perspectives

Taking a few steps back and reviewing the activities and initiatives taken since 1987 while watching the global ppm (Parts per Million) increasing annually, the outlook is not optimistic. The safe CO_2 limit is estimated to be 350 ppm. The atmospheric carbon dioxide measured in 2021 was 419 parts per million. Climate change is intensifying and affecting more and more livelihoods, while we are still discussing the fact that something must be done soon, what we should do and how to do it. The steps taken since 1987 have not been far-reaching enough. So much research, so much evidence, so many tools, with so little effect. One local friend maintains: "climate change has always happened during the past thousands of years. Sometimes fast, sometimes slow. But it is changing constantly, nevertheless". My point, shared by most people today, is this: if there is evidence—and there is overwhelming scientific evidence—that humanity has impacted the climate and the natural habitat adversely, we must, as a conscious species (homo sapiens), take responsibility, reverse the damage, and change our ways. There is no other way forward. Citizens are anxiously watching the follow-up on the new agreements on the national targets for CO_2 reduction. Will the politicians do what it takes to meet their set targets, or what will their delaying tactics or excuses be this time? The scepticism and mistrust are rampant due to years of broken promises in the context of not alleviating the crises and progressing sustainability. It has been said that the main excuse for not acting has been economic, as CO_2 reduction initiatives have been thought to burden a country's economic growth. In Denmark, the reason I heard was that the welfare state would suffer. However, the extraordinary financing that the government miraculously made available during the Covid crisis puts their economic reasons to shame. In addition, there is now a dramatic (however, welcome) change in attitude due to the Russian/Ukrainian war, which has highlighted the energy dependence on Russian fossil fuels.

To proceed with some degree of optimism, the next chapter will narrate three case stories that I researched for my PhD thesis. The cases demonstrate the degree to which sustainable development initiatives are possible in the current paradigm, what it takes and what factors must be present to achieve positive outcomes as well as the factors that hinder a transition process.

Readings and References

Bell S, Morse S (2003) Measuring sustainability: learning by doing. Earthscan, London

Bell S, Morse S (2008) Sustainability indicators: measuring the immesurable, 2nd edn. Earthscan, London

Benyus JM (2002) Biomimicry: innovation inspired by nature. Harper Perennial, New York

Birch EL, Lynch A (2012) Measuring U.S. sustainable development. In: Worldwatch Inst (ed) State of the world. Island Press, Washington, p 78

Blewitt J (2008) Understanding sustainable development. Earthscan, London, p 157

Blincoe KS (2020) Essential sustainability: from utopian theory to practical application. Dissertation, Kingston University. https://eprints.kingston.ac.uk/id/eprint/51146. Accessed 2020–2022

Braden G (2014) The turning point: creating resilience in a time of extremes. Hay House UK Ltd., London

Caradonna JL (2016) Sustainability: a history. Oxford University Press, New York

Circular Economy (2015) Ellen Macarthur Foundation. http://www.ellenmacarthurfoundation.org/circular-economy. Accessed 5 April 2015

Crawford M (2010) Creating a forest garden: working with nature to grow edible crops. Green Books, Totnes

CSR (2015) CSR. http://csr.dk/de-gode-penge-ligger-i-b%C3%A6redygtighed. Accessed 15 Sept 2015

Diamond J (2011) Collapse: how societies choose to fail or succeed, 2nd edn. Viking Penguin, New York
Doppelt B (2003) Overcoming the seven sustainability blunders. Syst Thinker ® 2 (14) no. 5
Ecological Footprint (2020) Global footprint network. https://www.footprintnetwork.org/our-work/earth-overshoot-day/. Accessed 9 May 2020
EEA 5-year Report (2018) European Environment Agency. https://www.eea.europa.eu/publications/environmental-indicator-report-2018. Accessed Jan 2019
Ehrlich PR, Holdren JP (1971) Impact of population growth. Sci New Ser 171(3977):1212–1217
Elkington J (1999) Cannibals with forks: the triple bottom line of 21st century business. Wiley, Hoboken
Farley HM, Smith ZA (2014) Sustainability: if it's everything, is it nothing? Routledge, Abingdon
Freerman RE (2010) Strategic management: a stakeholder approach, 2nd edn. Cambridge University Press, Edinburgh
Futerra (2011) The greenwash guide. Futerra Sustainability Communications, London
Gasparatos A (2010) Embedded value systems in sustainability assessment tools and their implications. J Environ Manag 91(1613–1622)
Goodall C (2007) How to live a low-carbon life. Earthscan, London
Gore A (2007) An inconvenient truth: the planetary emergency of global warming and what we can do about it. Bloomsbury, London
GRI (2015) The global reporting initiative. https://www.globalreporting.org/. Accessed 7 June 2015
Gross National Happiness Index (2015) Gross national happiness. https://www.grossnationalhappiness.com. Accessed 7 June 2015
Happy Planet Index (2015) Happy planet index. https://www.happyplanetindex.org/. Accessed 5 June 2015
Harper P (2013) Permaculture: the big rock candy mountain. The Land Issue, 14. Summer
Harvey M (2019) Utopia in the Anthropocene: a change plan for a sustainable and equitable world. Routledge, Abingdon
Henriques A, Richardson J (eds) (2004) The triple bottom line: assessing the sustainability of business and CSR. Earthscan, London
Hildebrandt S (2018) Verdensmålene - det vigtigste punkt på dagsordenen. KLS Pureprint, Copenhagen
Holdren JP (2018) A brief history of "IPAT" (impact = population x affluence x technology). J Popul Sustain 2(2)
Holmgren D (2011) Permaculture principles & pathways beyond sustainability, 2nd edn. Permanent Publications, UK
Hopkins R (2008) The transition handbook: from oil dependency to local resilience (transition guides). Green Books Ltd., Totnes
IPCC 6th Assessment Report (2022). https://www.ipcc.ch/assessment-report/ar6/. Accessed 1 March 2022
ISO (2019) International Organization for Standardization. https://www.iso.org/standard/37456.html. Accessed 5 July 2019
Jepson P, Blythe C (2020) Rewilding: the radical new science of ecological recovery. Icone Books Ltd., London
Jesinghaus J (2014) Bellagio principles for assessing sustainable development. In: Michalos AC (eds) Encyclopedia of quality of life and well-being research. Springer, Dordrecht. https://doi.org/10.1007/978-94-007-0753-5_167
Kibert J et al (2012) Working toward sustainability: ethical decision making in a technological world. Wiley, Hoboken
Kirby J, O'Keefe P, Timberlake L (eds) (1995) The earthscan reader in sustainable development. Earthscan Publications Ltd., London
Labelling (2020) OECD. https://www.oecd.org/env/policy-persectives-environmental-labelling-and-information-schemes.pdf. Accessed 15 May 2020

McDonough W, Braungart M (2002) Cradle to cradle: remaking the way we make things. North Point Press, New York
Natrass B, Altomare M (2001) The natural step for business, 2nd edn. New Society Publishers, Gabriola Island
One Planet Living (2019). https://www.bioregional.com/one-planet-living. Accessed 20 Nov 2019
Porritt J (2005) Capitalism as if the world matters. Earthscan, London
Porter M (2013) Social progress index. Guardian. http://www.theguardian.com/sustainable-business/michael-porter-health-happiness-index. Accessed 1 May 2013
Pre-sustainability (2020) Quantifying sustainability. https://www.pre-sustainability.com/news/2020-the-year-of-the-perfectly-sustainable-world. Accessed 5 Dec 2019
Princen T, Maniates M, Conca K (eds) (2002) Confronting consumption. MIT Press, Cambridge
Rio Declaration (1992) UN. https://www.un.org/en/conferences/environment/rio1992. Accessed May 2019, 5 Dec 2021
Robert K-H (2002) The natural step: seeding a quiet revolution. New Society Publishers, Gabriola Island
Roddick A (2001) Business as usual: the triumph of Anita Roddick. Thorsons, UK
Rosling H, Rosling O, Ronnlund AR (2018) Factfulness. Sceptre, London
Sala S, Ciuffo B, Nijkamp P (2015) A systemic framework for sustainability assessment. Ecol Econ 119(314–325):1
Sharpe A (1998) A survey of indicators of economic and social well-being background. Canadian Policy Research Networks, Ottawa
SPI Social Progress Index. https://www.socialprogress.org. Accessed 15 Jan 2019
Sustainable Development Goals (2015) UN. http://www.un.org/sustainabledevelopment/sustainable-development-goals/. Accessed Dec 2015, 10 Jan 2022
The Brundtland Report (1987) WCED. https://sustainabledevelopment.un.org/content/documents/5987our-common-future.pdf. Accessed 9 Sept 2015, 14 Dec 2021
Wackernagel M, Rees W (1996) Our ecological footprint: reducing human impact on the earth. New Society Publishers, Gabriala Island
Wellbeing Economy Governments' Partnership (2022) Wellbeing economy alliance. https://www.weall.org/wego. Accessed 24 April 2022
World Economic Forum (2022). https://www.weforum.org/. Accessed 28 April 2022
WWF (2022) World Wildlife Fund. https://www.worldwildlife.org

Chapter 6
Three Transition Stories

Abstract This chapter presents the narratives of the three cases: the Danish island of Samsø, Brahma Kumaris, an international faith community, and Novozymes, a Danish multinational company specialising in the production of enzymes. The chapter is an updated version of the case studies in the thesis. The descriptions of the cases are based on data gained from the interviews I held with the leading teams. The main focus is on the specific process, that each case engaged in, including the successes and the barriers encountered. The three cases are very different. The diversity comprises culture, identity, focus, vision, context, size, and the reasons for engaging in sustainability implementation. The chapter demonstrates sustainable development in action and what was and can be achieved in the current framework, regardless of whether the undertaking is an island, a faith community, a corporation, or similar entities. It is due to initiatives such as these that the sustainability agenda has made progress since 1987. The positive perspective the chapter offers is the knowledge that necessary changes can be made despite barriers, adversity, and difficulties, provided the motivation and determination are present.

6.1 Transition Processes

> Never doubt that a small group of thoughtful, committed citizens can change the world. Indeed, it is the only thing that ever has (Margaret Mead 1901–1978).

The three stories describe the level of achievement that is possible when individual entities embark on a journey of profound change. A high level of sustainability can be achieved through individual initiatives when they are specific, set in one particular time frame and place, and with a particular focus. Apart from inspiring like-minded actors through their visible and concrete results and positive knock-on effects, they can serve as pilot studies for further larger-scale initiatives. The positive and pioneering results that these cases have achieved have given them national and international recognition. One has become a role model for small-scale communities, municipalities, and islands around the world. Another became a role model for corporate engagement with sustainability within a UN context and participated in the

K. Blincoe, *Achieving Sustainability: The Ultimate Human Challenge*,
https://doi.org/10.1007/978-3-031-10023-9_6

development of the 17 SDGs. The third case became an accredited observer organisation to the UN Framework on Climate Change and has inspired and led sustainability initiatives within the UN COPs for several years. The cases are three out of numerous attempts to implement sustainable development globally over the last three and a half decades. Their merit and importance, as with the utopian experiments, is the fact that they have a significant impact on promoting transition processes, locally and globally, as an inspiration and guidance drawn from real life. Theory, assumptions, and beliefs can take us on a journey towards a goal. However, experience based on reality is vitally important when planning processes that are deeply transformative and impactful, such as a transition to sustainability.

For the research, I chose a small island community, a large faith organisation, and a multinational company. It was essential to select diverse examples that were engaged in similar processes to be able to compare similarities and differences viewed from different perspectives. Another consideration when choosing such diversity was, of course, the fact that everyone has to engage in the transition, and that the process of change, therefore, has to be possible for all actors in society. The key resulting similarities and differences are discussed in Chap. 7. This chapter tells their fascinating stories using as far as possible the same tone and phrases that emerged from the interviews. There are many quotes or statements throughout the narratives. These are from the interviews. The interviewees did not want to be quoted directly.

Case Story One: The Island of Samsø Samsø is an island of the east coast of Jutland in Denmark. It covers an area of 114 km^2. There are 3,700 citizens on the island gathered in one municipality with several villages. The citizens consist of farmers, small producers of food products, and owners or employees of small service, support and tourist industries. The sustainability project on Samsø was initiated through a Danish political initiative in the form of a competition to create a renewable energy island, emanating from the Kyoto discussions in 1997 (Sect. 2.4). Samsø won the competition and was given seed money to start the process.

Case Two: Brahma Kumaris Brahma Kumaris is a faith community/charity with a focus on world peace. The headquarters are situated in Rajasthan, India, and have approximately 4,500 regional offices in 110 countries on all continents. The organisation has approximately 1 million followers. The local centres are independent organisations or charities collated under the Brahma Kumaris umbrella. Brahma Kumaris is an official international non-governmental organization (NGO), accredited by the United Nations. Their work is educational and charity based, focussing on creating world peace through inner personal transformation, mainly through meditation practices.

Case Three: Novozymes Novozymes a Danish corporation with an international reach. The headquarters are based in Denmark. Novozymes specializes in enzyme production for approximately 700 products that are used in and around the home daily, such as washing powders, bread, beer, juices, clothes, paints, and cleaning agents. Novozymes has offices in 130 countries and had at the time of the interview approx. 6,500 employees worldwide. The company is quoted on the stock exchange and has been an active collaborator with the UN on the development of the 17 SDGs for a number of years.

6.2 Case Story One: The Island of Samsø

Our goal of achieving 100% renewable energy sources in 10 years we achieved in 7 years.

Samsø is a Danish island positioned in the North Sea, called Kattegat to the east (approx.15 km) of the mainland, Jutland, and south of Norway. It covers an area of 114 km^2. The municipality of Samsø has approx. 3,700 inhabitants called Samsings. The island has, from older times, been the home of small-sized farms, mainly growing potatoes (the island's potatoes are famous in Denmark) and asparagus. The island had several cooperatives, established in the nineteenth century when the cooperative movement swept the country. The cooperatives included dairies, local shops, and later a slaughterhouse for meat products sold to the mainland. People were employed on the farms and in cooperatives. In the early twentieth century, there were approx. 7,300 inhabitants on the island. This number slowly decreased due to a lack of educational possibilities and employment opportunities. However, the mild climate around the island promoted a growing tourist industry, which currently constitutes the main form of employment and income generation on par with agriculture and renewable energy expertise.

A Renewable Energy Island

In 1997 Samsø won a government competition to become a model renewable energy community. Svend Auken, the then Danish environment minister, promised in Kyoto, Japan, to cut CO_2 emissions in Denmark by 20%. He announced that he would prove it was possible to create a community where the energy was 100% renewable. The minister organised a competition to develop renewable energy islands in Denmark. Samsø won the government competition. The island was until then dependent on oil and coal imported from the mainland and there was no district heating system.

Should we go for it?

This was the question for the islanders as the project was collaborative and included the island community. A public meeting was organised, and teachers, farmers, shopkeepers, smiths, carpenters, and others turned up. The organisers said: "We would like to embark on this because of the chance to become the first Danish energy island". The people present agreed that it would be interesting to be part of the experiment. They formulated a declaration and created an organisation with a steering committee, and got the local mayor involved. He supported the project from the start seeing this as a way to create new jobs, be at the forefront of development, and open up for new ideas to strengthen the survival prospects for the island.

The environment minister did not give any support apart from the title of becoming a "Danish Energy Island". The community had to find financing to develop the plan themselves. They wrote an application to the Danish Energy Agency and received 200,000 DDK (€27.027) in funding to write the master plan together with several collaborators. The long process to develop a renewable energy island had begun.

> The island has a tendency to be a small world, different from other areas on the mainland. An island is more special - you have to take the ferry.

Today most of Samsø's electricity comes from wind power, and 75% of heating is sourced from photovoltaic cells and biomass energy. Some vehicles are powered by locally produced biofuels (rapeseed), and there is an increase in the purchase of electric cars. The Energy Academy was established on the island with a visitor education centre that hosts approximately 4–6,000 visitors a year. They are called *energy tourists* and come to learn about the renewable energy project. Samsø receives in the region of 100,000 tourists annually, providing the second largest income after farming. The academy is currently working with the municipality to make Samsø 100% fossil fuel free and is, together with the local community developing the next phase of greening the island. There is a focus on developing a sustainable transportation system using renewable energy forms as well as a focus on developing organic farming producing a variety of vegetables, grains, and fruits, meat, and livestock products. There is an increase in dairies, breweries, restaurants, and cafés—as well as permaculture, foraging, and forest garden experiments. Several master plans are in the pipeline to extend organic food production and education in renewable energy systems. There is a wish to create more jobs and stimulate new settlements on the island, secure its future, and make it a role model for sustainable communities nationally and internationally.

Vision and Mission

> To get the community on board to make it happen.

The primary official and public vision in 1997 was to make Samsø a renewable energy island over 10 years. However, there were 'sub-visions', which were as important, if not more important than the main vision. These included reviving and revitalising Samsø through income-generating activities, reactivating old businesses, and developing new areas of work. It was important for the island to become independent of fossil fuels and fluctuating oil prices at the same time as energising and empowering the local population. "Reduction of CO_2 emissions were very low on the list of reasons", was one of the comments. The survival of the island was the priority, and for the islanders to participate and be able to decide for themselves as a community. Another of the island's goals was to find a way to keep the young generation on Samsø: "They leave the island to get an education and not many people come back. Only 10% return to the island after their education", it was said.

It became clear to the initiators at the start of the project that the key mission to achieve the goals was to get local community support. This included engaging in activities to replace the existing fossil fuel-based energy- and heating systems and move towards alternative systems. It was felt that unless local people became involved, the project might easily fail. Subsequently, many meetings, working groups, and interest groups were organised, and over 100 local citizens became directly involved in the process, while many more supported the initiative.

Philosophies and Values

The philosophy underpinning the work was based on a hands-on, bottom-up approach. The islanders stressed that there were both practical and idealistic reasons for starting the project. For it to happen, the process had to be hands-on. A "survival" kit was developed to produce jobs, create local independence, stop the dependency on fossil fuels and create new possibilities for the development of Samsø. This meant involving people, both young and old and generating practical tasks and methods for them to engage with. The philosophy was centered on a deep understanding and respect for the islanders, their traditions, way of life, and love of the island, as well as an understanding of the island's social structure in terms of the decision-making process.

The project, it was stated, was built on the values of the old agricultural community—such as the ancient Viking community, where wise and mindful use of resources, protection of the community, and trust were essential values. Samsø rests on an old island hierarchy, where collaboration and social relationships are core elements—they are at the heart of the community.

Their core values include trust, respect, wisdom, empowerment, transparency, cooperation, and collaboration. The power of the combination of these was demonstrated by the people leading the project. The current CEO of the Energy Academy and leader of the initiative is a seven-generation farmer of Samsø. He was well-known and respected by the local community, down-to-earth, had great communication skills, was clever, and had a good sense of humour. Combined with respect for the local community and love of the island, these were qualities that made people listen. The leader was supported by an organising team with similar qualities and who were part of driving the process to get the initiative up, running, and progressing.

Sustainability

> It is about being confronted with all the shadow sides of the welfare society.

The philosophical view was that sustainability, from a universal perspective, was about living in balance with the earth and its resources, whereas sustainability in the island context was about community, collaboration, jobs, social interaction, and empowerment. Sustainability, it was said, constituted a choice, a consciousness around the choices one makes and the consequences it causes. It was about downsizing and planning a balanced future, assessing the available resources, and utilising them effectively in the short as well as in the long run. Sustainability was not only about self-sufficiency but also about freedom, and freedom in the interaction with people. It was about having a core—a soul, and about continuously "sharpening the sword like the Samurai warrior". Sustainability was about choices. "Do you need this? Do you need that? How can we improve and renew while caring for the resources we have and our stewardship of them?". It was about being able to see seven generations ahead. The farmers did that in the past, as it was cultural heritage and part of their DNA, where balance was an integrated part of their livelihoods.

Sustainability, it was also said, should be the rule and not the exception to the rule. Having a pragmatic view on ecology, nature, and spirituality—was where transformation should start, and it was a bottom-up movement, not a top-down scientific process. The sustainability paradigm was about: Does your life make sense? Your work? Your priorities? Your choices? The holistic aspect lies in the detail. If the pieces do not fit together and create a balanced whole, there will be a collapse, it was said. There was a view that if it were known precisely what kind of expertise constituted sustainability, it would be possible to define, evaluate and acknowledge the concept, and create a relevant balance between the inputs from experts, facilitators, and practitioners.

Sustainability in the Samsø context was seen as both a locally orientated and hands-on concept. For example, farmers should not be regarded as bad people because they did not grow organic crops. Instead, sustainability should be about talking to them and discussing the issues, such as how could fertilisers be produced on the island and not be imported. How could carbon be captured from the air and used as a fertiliser in the fields How could the farmers grow organic crops and make a profit? Sustainability encompassed local, collaborative factors, which meant reaching a joint agreement on the elements to be implemented, what process and methods to use, and doing something about it.

Reflections on sustainable development also included the idea of developing a closer link between the urban and rural areas. The rural regions were producing resources such as food for the cities. The cities were seen by the islanders to give nothing in return, apart from pollution. "Carbon is exported to the cities; it is digested and burnt in generators for cheap heating for houses and nothing is given back". Not a good bargain for the rural areas. "We want the shit back in bags to use it as compost, so to speak". It was about creating a circular economy and a reciprocal resource system between the city and the surrounding countryside so that both parties could benefit.

Spirituality was considered an integral part of sustainability. It existed on Samsø in a different form from the usual understanding of spirituality. The community was not specifically religious or spiritual. But there were key elements present, like a belief in loyalty and love. Belief in a better way of life, a better island, a better world—an inner wish to change society. Loyalty towards what is, what was, and loyalty towards the island and islanders. There existed a deep love for the island, its families, and its history.

6.3 Process and Key Success Factors

There was a lot of scepticism at the start of the initiative in 1997. When the competition was won, the mayor was interviewed on local TV and was asked the above question. He did not know what to answer. No one had thought about how they were going to realise the project. The renewable energy island was a nice title, and the expectation was that ample funding would be available. As this turned out not to be the case most of the locals thought: "we have heard this before......about change and initiatives that come in from the outside and end up in nothing". So, the attitude was that if they just kept quiet and waited without doing anything, the "outsiders" would lose interest, and life would carry on as before. The local community had previously seen many plans proposed for the municipality developed by outside consultancy firms and knew that many of these were never completed. The first and key job of the leading team was, therefore, to work with raising awareness of the project and its potential and make the islanders aware of the ownership possibilities and the what's-in-it-for-me elements. People were stressed about the increasing unemployment and an uncertain future for the island. So, to raise interest and alleviate some of the worries, information-sharing seminars were organised that were focused on openness and transparency. This was the first step. The second step was to work out a master plan and make it practical, which they did.

One method used repeatedly in the many meetings participated by a cross-section of the inhabitants was making detailed preparations. The organisers prepared the plans to be presented in minute detail. They spoke with key local people, who were going to attend, and explained the plans to them prior to the meeting. The organisers knew that if these people backed the plans, the other people present were more likely to go along as well.

Cooperation and Collaboration A major factor in the success of the project was the cooperation and collaboration at all levels. The management team said: "We adopted the holistic approach, kept the old-time village structure, where the big farmer and the little farmer had to work together. We created many meetings in the villages with the village organiser. We brought the coffee and cakes and proposals to their premises. We had open discussions where people could say, "I don't believe in these things". It was ok to be negative. We said: "Let's see what we can do, and you can make up your mind now or later, as long as you agree that we should try". They all said, "yes, let's try; we can always reject it".

The project leader explained: "To begin with, I was very enthusiastically talking about changing the world, but the locals rejected this completely. I had to go home, reconsider and ask what went wrong there. There had to be some carrots. I asked the engineers to break up the project into smaller bits, i.e. we need so much manpower, so much energy, and so on. Then I had meetings with the bank on the finances so that we could put the paper with the figures on the table and say: This is the cost. It will cost you so much per year to do this project". The project leaders had discovered that they could save more than 20% on overall energy costs on the island and spoke to the bank. The bank liked the project because it made financial sense. The house owners liked it because they could save on the price of energy. Everyone was able to see the what's-in-it-for-me aspects and the project progressed.

Communication The project leaders used a variety of communication channels and skills to spread the initiative and get as many people involved as possible. They asked the local craftsmen, small businesses, and service providers such as smiths, plumbers, painters, carpenters, and electricians to promote the project whenever they went out to do repairs on the island. However, it became apparent that even if they agreed to do this, nothing happened. The project leaders then decided to offer courses and training in renewable energy skills to small businesses. This produced the desired results as the businesses were able to sell their normal services and offer new skills within the renewable energy field, such as installing solar panels, mass ovens, insulation, and carrying out energy checks.

Business-Like Approach The experience was that even if the work had to be locally orientated and hands-on, the approach had to be prepared and presented professionally. The business-like approach was important for people to take the work seriously, and many hours were spent by the leadership team to make sure there was a focus on the details. The team did the practical work to make it easy for the islanders to participate. When the first wind installations were in place, and they worked, it attracted more interest from the locals. More people wanted to invest, and the project leaders helped to organise individual loans to help them overcome the fear of going to the bank and asking for money. The process worked and became successful. This was also due to the fact that if someone did not trust the process or the information presented regarding plans, facts, or figures, informal meetings would be set up, and the issues would be dealt with over a cup of coffee in people's homes or workplaces. "We needed everybody on board, so we addressed one village at a time—in all 23 villages. We estimated that 60–70% of all households would be interested in district heating based on renewables. Some houses were outside the limits of district heating, and they needed help to install their individualised heating systems".

Additional Factors

> We had as much as possible owned by people and structured by people.

The preliminary work and preparation activities by the project leaders were a critical factor in the success of the initiative, it was stated. It created a sense of being looked after and respected, as it was easy for anxious people to generate myths and gossip, and invent reasons not to participate. The attention to detail and the ease with which people could become engaged, as well as the successful first installments of the new energy systems, helped the process along and surrounded it with positiveness. Another aspect that helped further the interest was the increase in oil prices. Even though the national prediction at that time had been a decrease in the price of oil, this did not happen. Instead, there was an increase in the oil price, and this influenced the project in a positive way. People began to realise that there was an added value in investing in green initiatives and reducing dependence on fossil fuels, thereby creating greater potential for the island.

> You could save money and be protecting the environment and still have a good house, where you don't lower the comfort.

The positive returns on investments from the wind turbines were very significant. One citizen who bought shares in the windmills said: "I did not have any money and had to borrow it, and my parents had to countersign so that I could borrow the money. I bought two shares for 30,000 DDK (€4,030) each. They have been paid back, and I now earn 1,000 DDK (€134) a month on the shares. I paid the loan back over eight years. It was a good investment for the people on Samsø, who bought shares". There were harsh and fierce discussions with the farmers, as they initially wanted to own the wind turbines or the leases on their land. However, many farmers who invested in the project said afterwards: "It was a good investment".

The renewable energy initiative created a change in attitudes. The mindset changed. Instead of buying vacations, going to "Thailand on holidays, and buying red wine", many islanders started to invest in renewable energy, wind turbine shares, insulating their homes, and currently, many are buying electric cars. It became a competition to engage in something, not just to save the planet but also to save money. This helped create a better local economy in Samsø. The average income was the 4th or 5th smallest in Denmark. However, the islanders succeeded in building the biggest offshore wind turbine park in the world in 2003, and Samsø became a renewable energy island in 2007. They were and are very proud of that.

Other gains followed in the wake of the installation of the new energy systems. The price of electricity remained the same, but the profit now remained on the island, as the islanders created their own electricity company. Households with solar cells on the roofs generated their own electricity and did not have to buy from the grid. The awareness of consumption and attention to energy saving increased. The average electricity consumption on the island decreased, even if the change to heat pumps increased. In comparison to the mainland, this was another positive outcome, as the mainland had an average increase of 6–7% in energy consumption p.a. Collaborations between the municipality and local businesses on Samsø were set up to look at what more could be done, based on the experience gained. They wanted to work out the best solution for each new initiative and calculate the consequences and the what's-in-it-for-me gains before taking action.

The following is a listed summary of what worked well in the process.

- Joint and collaborative leadership.
- Hardcore strategies.
- Elaborate and detailed planning.
- Good, clear, and ample communication.
- Attention to detail.
- Trust between the stakeholders.
- Transparency, loyalty, inclusivity, connectedness.
- Ability to listen.
- Honest talk.
- Sharing of and respect for cultural and traditional values.
- Sharing of the good story.
- No backstabbing.

6.4 Difficulties and Barriers

The process was not just "smooth sailing", it was said. There were difficult periods and many hurdles. Communicating the initiative so that people could see, feel and understand what the project was about was difficult. Secondly there was a lack of people, who wanted to participate without the view of gaining financially or having the what's-in-it-for-me attitude.

Three main factors drove the initiative: the heart, the mind, and the wallet. When asked what came first, the answer was: "the wallet". There was a sense that the project could have been more diverse. It was to some degree felt that the long view was absent in the processes, and that there was too strong a focus on purely economic aspects. The project was carried out and led by a group of strong-minded people who worked hard to generate success. It was a tough process and took its toll on people's health. Other negative issues included some islanders gaining more financially than others. This was at times a contentious issue as there was a view that if the plan had not been broken down into short-term results with financial gains for the locals, they might not have become involved, and the project would have suffered from lack of investment. There was a keenness to get acknowledgement from central government in Copenhagen and prove that the project could succeed, as well as a wish to gain international recognition. This caused some tension as it created pressure both from the outside as well as from within the community to generate results.

The following is a summary of some of the problems encountered:

- Big changes are complicated in small, conservative communities like Samsø.
- Once a project has been promoted in a manner where people reject it, it is very difficult to course-correct.
- "You have to be very careful how you do things. You can't be too open-minded and be this happy fellow who is advocating green transition, because people will see you as an irresponsible Copenhagen hippie".
- Getting the wrong people on board can cause problems.
- People might support activities at meetings, but behind "the scenes" be talking badly about the project.
- People are naturally conservative, and engage in small talk and gossip. This can harm the process.

> Never give up, always go for solutions. Always looking for the best solution.

The way the above barriers and problems were overcome included the project leaders listening carefully and mindfully to the local community. They listened to the background gossip to discover what was going on, as many people were caught up in realising their personal ambitions. From time to time, the development process had to be slowed down or even stopped while changes took place. The project leaders talked and listened to key community members to hear what they thought, the problems they saw, and how to move the process forward. Without these on board, "we will have problems forever. That is a fact". "Some projects did not succeed, and I still don't really understand why, but I believe we did not get the right people on board from the beginning".

External factors caused problems from time to time. For example, the wind turbine companies did not understand why it was taking such a long time to get everyone on board. However, the locals demanded that the initiative should respect their wishes; otherwise they would not lend the necessary support. "It may not be the most rational way, but it is the only way here in this community. It is either this or nothing". The difference between the two attitudes lay in the fact that the renewable energy suppliers "were sitting behind their desks in their offices and were not on the ground", and there was a disconnect between them, which did not generate the desired results. Other outside factors had a negative impact, such as changes in government policy during the process. Budgets and subsidies were cut. Luckily there were signed agreements with the government for the main part of the work on the island. Some projects, however, were terminated due to policy changes, which was not good for progress. Ten years of political planning was important, but with policy changes every two years, people got weary and started to worry: "What is going to happen to my budget and what about taxes?" Transparency and honesty were the only way these issues could be dealt with. Honest and clear communication.

Reflections

Looking back over the past 17 years, the people who were interviewed reflected that were they to start the initiative today, they would make use of the same methods. However, the external parameters did change. Since 1997 national budgets have been adjusted, and green policies have been put in place. Green development was no longer an idealistic project but became a market-driven issue. The people of Samsø, therefore, had to find new goalposts for developing the island and decide on the next step. They had learnt that to succeed, a certain size of the community worked best and that if their community increased in size, they would have to put new thinking in place as the coherence might otherwise break up. They had learnt the importance of taking smaller steps at a time and not having working groups that were too large. Most of all, to keep asking people what they were comfortable with and being good at listening.

Another important point was that the story mattered, the good story about what got them there, the story about becoming a role model. It was stated that due to a number of factors, the project became more impactful than expected—there was a synergy or symbiosis, an allowance for emergence to happen. The timing was right, the leaders were the right leaders, and the direction was the right direction. Locality, activity, and mentality joined together and created a mind-shift that caused the goal to be achieved in seven rather than the estimated ten years. This mind-shift also attracted new players/people to the island, who contributed to the development of new opportunities. Clear communication between the government in Copenhagen and the people of Samsø was a final important factor. The islanders felt there were too many regulations, too many laws that controlled local communities, and too few bridges to close the gap between the rural and urban communities. Their wish was to have more responsibility locally and more decentralisation.

The renewable energy project on Samsø was completed three years earlier than estimated. The islanders decided to build on the success they achieved and have continued to green their island. From focusing on electricity and heating, they have turned their attention to transportation, water- and waste management in terms of up-cycling, recycling, and water treatment processes. Until that time, the island had paid to get rid of waste. Now they wanted to utilise it as a resource. Food production was looked at. The Samsø farmers were paying a higher price to transport their products than farmers on the mainland, but they realised that they could compete on quality and their brand. An association was established to increase the opportunities for organic farmers by collecting funds to buy farms to grow organic crops. Young families can now lease these farms for 2–3 years and after that be allowed to buy them. Discussions were also centered around social topics such as new values, health, security, demographic changes, and well-being as the next targets for sustainable development for the island.

The plan is for Samsø to be 100% fossil-free ahead of the rest of Denmark and the EU and embark on implementing a circular economy on the island as well as working to meet the 17 Sustainable Development Goals. The islanders feel that they have a past, a present, and a future and want to build on and utilise the experiences and lessons learnt. The island has become world famous because of this achievement. The renewable energy work was awarded by the UN at COP 26 in 2021, when Samsø received the UN Climate Leader Award, which is part of the climate action programme.

The Samsø identity has changed. People have become empowered, strong as well as proud of what they have achieved. There is a calling for the island, a pull towards the future based on a new understanding of their shared set of values, philosophy, methodology, and process. The islanders have learnt what stability, consistency, strength, and community mean. What Samsø has achieved can, in many ways, be said to be exemplary, given the context in which the island previously operated. The case provides valuable lessons and can act as a model to those who want to develop sustainable intentional or smaller mainstream communities. Careful scrutiny of the Samsø story and process can provide clues and critical elements to embark on larger-scale initiatives. Here again, there is a direct link to utopian initiatives relating to the drive for change for a better life and a better environment. The important difference is that Samsø achieved its goals within the existing societal framework, still being part of the mainstream.

The mood and hope for the future are positive. Samsø has become a flagship of Denmark's sustainable development process. The island has shown that with dedication, wisdom, and people-orientated leadership and skills, a great deal can be done by individual communities to further sustainable development initiatives.

6.5 Case Two: Brahma Kumaris World Spiritual Organisation

> Through a variety of activities and partnerships, the organisation promotes spiritual understanding, universal values and leadership with integrity towards a better world (Brahma Kumaris 2019).

Brahma Kumaris (BK) is a faith organisation and an international NGO affiliated with the UN. Brahma Kumaris was established in 1937 as a spiritual school mainly for women in India. Today, it is an organisation that is dedicated to peace, personal transformation, and "world renewal" with a focus on teaching Raja Yoga meditation. Raja Yoga is a meditation method that focuses on the soul or soul energy to help "individuals transform their perspective of the world from material to spiritual". "It supports the cultivation of a deep collective consciousness of peace, and the individual dignity of each soul".

The founder of Brahma Kumaris was a wealthy businessman, Dada Lekhraj, in Hyderabad (then India, now Pakistan). At the age of 60, he started to experience visions of a new and peaceful world and felt inspiration coming from God, giving him the task to set up a spiritual school that had to be run by women. Brahma Kumaris means "daughters of Brahma". It was the first school for women in India and was controversial at that time in the male-dominated Indian society. There was, therefore, substantial resistance to the project from mainstream society, but the school grew and is today an international NGO with approx. one million followers worldwide.

Respect, equality, and humility

In the beginning, the school educated women to become teachers and leaders based on spiritual principles, with the aim to promote world peace. There was a belief that if a school to promote world peace was to succeed, it had to be developed and run by women. However, Brahma Kumaris is not a woman-only organisation. It is managed and run by women, "elder sisters", but men are represented in middle management and as leaders of local and regional centres as well as teachers. Brahma Kumaris has, since its inception, stressed the respect for equality between men and women. There are currently 60% women and 40% men in the organisation. The leadership model is based on "respect, equality and humility".

Many of the original group of 300 members who started the school and organisation became teachers, and some left India to set up centres in different countries. There was no master plan or strategy in terms of expansion. The organisation grew organically—and is still growing. Only on the African continent did Brahma Kumaris create a plan, as they felt there was a particular need to help the African countries.

Brahma Kumaris spiritual headquarters, Madhuban, are based on and around Mount Abu in the Northwest state of Rajasthan in India. BK is a worldwide community with a large infrastructure and has over 4,500 centres, most of them in India. Centres are located in 110 countries throughout the world, and Brahma Kumaris has a wide reach in the communities surrounding the centres. The service it provides is not a religion but promotes and teaches a belief system, philosophy, and a way of life that focuses on the world and inner peace and provides an educational platform for adults. The members or followers include all races and religious denominations.

Brahma Kumaris is internationally recognized at many levels for its work. It's an international non-governmental organisation (NGO) of the United Nations, which has general consultative status with the Economic and Social Council (ECOSOC). It is also affiliated to the UN Department of Public Information (DPI) and is an accredited observer organisation to the United Nations Framework Convention on Climate Change (UNFCCC), Biological Diversity (CBD), Combat Desertification (UNCCD) and the United Nations Environment Program (UNEP). The website states: "In promoting the purpose and principles of the United Nations, the spiritual trajectory of awareness, attitude, vision, and action is used by Brahma Kumaris in the areas of the Millennium Development Goals, Climate Change, Food Crisis, Gender Equality, Global Public Health, Humanitarian Emergencies, Human Rights, Women, Children, Youth, International Decades and Days" (BK 2020).

Vision and Mission

> I create my world from the inside out.
> The thought energy I send out will shape my life.

The Brahma Kumaris vision is peace and happiness on earth. The vision has remained the same since 1937 when the organisation was established. The mission of Brahma Kumaris is to create the transition to a better and peaceful world through educational activities and being a role model for the surrounding environments. The belief is that a world of peace and perfection exists and that everyone should be invited to join the process of getting there. However, creating a world of peace and perfection requires change from within. Once people change other actions will follow, transition will occur, and the problems facing us can be addressed and solved.

The core of the teaching in Brahma Kumaris is Raja Yoga. Raja Yoga is a meditation practice that addresses inner change through connecting with oneself as a soul, and with what it calls the supreme light (God), and to share with others. The teachings are communicated through daily mediations, talks, seminars, courses, workshops, conferences, and the internet. All courses, seminars, workshops, and meditations are free and regarded as community service.

Brahma Kumaris centres organise activities both in and outside their local communities. The teachings promote holistic health and agriculture to citizens in many areas of society. The centres also engage in social work. They look after the poor and street children, help to empower women, work with prisoners, and offer daily meditation and courses in self-development and personal change. In addition, the organisation has, for approximately 25 years, been engaged in research and development, as well as the implementation of renewable energy projects. BK is actively engaged in and with international climate and environment conferences, where their role is to promote lifestyle changes to protect the future of the planet.

Near the headquarters on Mount Abu, the organisation established in the late 80s a hospital, the Global Hospital and Research Centre. It became registered as a charitable trust in 1989 and was supported by various charities throughout the world. The foundation promoted and researched "a value-based model of modern health care, offering improved understanding in response to suffering and encouraging the development of positive thoughts and attitudes" (Global Hospital and Research Centre 2022). At the hospital are several departments using a variety of medical systems and practices, for example, the Western traditional, the ayurvedic and homeopathy-based systems. Yoga and meditation are also taught at the hospital, and the treatment of patients is holistic, taking into account the whole person, such as the physical, mental, emotional, and spiritual. Treatments are free for poor people. There is a mobile outreach unit which treats the poor in local communities in remote areas.

Philosophy and Values

> I need to practice my values, no matter what the response is out there.

The philosophy and values have remained the same since 1937. The teaching has changed with a slight adaptation towards the zeitgeist (the mood of a period). The main philosophy includes teachings about the soul and supreme soul. Brahma Kumaris practices and teaches Raja Yoga meditation which relaxes the mind and nurtures a healthy balance between the inner and outer worlds. BK believes that the soul essentially contains the elements of peace, love, happiness, purity, and power. Those are innate qualities. There are also core values. They include tolerance, peace, purity, compassion, and understanding. There are other non-core values, such as environmental concern, simplicity, responsibility, and care. One must care for everything, including things, plants, and animals. The essence of life is caring. Simplicity is about feeling contented. A simple life does not mean that one can not live a full life, but it is the belief that when there is contentment, simplicity becomes a natural way of life. One chooses a simple lifestyle rather than compensate for dissatisfaction with distractions.

Brahma Kumaris believes in destiny. A person is destined to play the role of her/his life (as an actor) in the play of karma = action and reaction. Brahma Kumaris believes in the re-incarnation of the soul. The way a person's life unfolds is a result of previous actions, good or bad. Past karma (actions) determines the way people live their lives today. The more good actions you engage in, the easier your next life becomes. BK also believes that humanity is currently going through a cyclical transition, as the old systems are not functioning anymore. The climate, environment and economic systems are failing. To solve a problem, the tendency is to invent a new technology, but often this creates further problems. Technology by itself is not the solution. A paradigm shift is imminent and needed. The shift has already begun but is not mainstream as yet. However, here and there, people are changing their ways of being and their lives. The old system is breaking down, and the new is already present as "a seed growing up from the ground".

There is no membership involved in becoming part of Brahma Kumaris. It was and still is a voluntary organisation, and people who give their time and skill do not receive a salary.

The daily routine and principles for living include:

- "Meditation early in the morning.
- Taking part in the morning classes.
- Read/study the Indian texts (Murli).
- Entertaining a vegetarian diet.
- Living in celibacy.
- Avoiding drugs of any kind.
- Optimising the spiritual efforts".

The framework and principles for becoming a student of the Brahma Kumaris community can be challenging and do become problematic for participants from time to time. However, they can leave the organisation, if and when they wish, as the essence of the community is based on voluntary commitment. Each centre is financially independent and is registered in the local country. The centre uses voluntary local help and support. In addition, the centre also gets support from the main Brahma Kumaris regional centres and headquarters in India, mainly through advisory services and the availability of teachers.

Income is generated from voluntary donations. This principle has been kept as a founding principle of the organisation. According to Brahma Kumaris, the system works because it is seen as part of a healthy growth pattern. You receive as much as you give, and this is a healthy process, because it aligns with spiritual vibration and because it is seen as being just. Brahma Kumaris believes that when income is given from the heart, it is more valuable than something given with resentment. This generates *happy* money that can be utilised more positively. Funding is given with good wishes, which again is considered as a kind of a special *fertiliser*. What is subsequently created with this type of income is seen as whole and containing positive vibrations.

Brahma Kumaris believes in healthy and organic growth, and the organisation is careful not to get into debt, borrow money or *put pay-day in the future*. Healthy growth is based on the availability of *good* money and a clear idea of what makes sense. That in itself is believed to have a healing effect on the world. This belief has developed and grown with the elder sisters at the helm. They developed this methodology for healthy growth. This meant that the organisation expands sometimes at a slow and sometimes at a fast rate. In recent years, the demand and need for spirituality have grown, and Brahma Kumaris is preparing to increase its capacity to embrace this.

Sustainability

Brahma Kumaris believes that sustainability is about long-lastingness and that the concept can be compared to "how a mother looks after a child, or a gardener looks after the garden"—"in that sense, it is very close to spirituality". Sustainability is viewed as a long-term commitment, not a temporary or quick fix, but something which needs to be nurtured over a long time. This commitment requires tolerance, patience, and changing the self, and these elements are seen as inherent values of sustainability. Some of the thoughts that were expressed included. The view is that there is a lack of new visions, such as those developed by the Club of Rome and Limits to Growth, and the call is to develop new ideas supported by clean technologies and a change of lifestyle. There is a need to collectively change people's attitudes and awareness; then, it is believed, a healing process of the planet can start (Club of Rome 2019).

Personal Sustainability

At the core of sustainability lies personal sustainability. Meditation and spiritual practices must be undertaken ongoingly and must be sustained over many years. A variety of focused meditations are recommended, for example, on spreading peace and harmony on the earth and meditations on the natural elements. Spirituality is regarded as an inherent part of sustainability. It is seen as a way to be supported from the inside as a help to overcome or embrace short-term or long-term obstacles and to live a life of simplicity.

> Nature brings out your true nature.

The natural environment is important in Brahma Kumaris. Walks in nature and time spent in nature are encouraged as well as outdoor meditation. "Nature brings out your true nature", the yogis say. They believe in preserving and not harming nature. The sun has a mythological meaning and is a deity called Surya. To cook directly by the sun is highly respected, and is considered to have health benefits, hence the installation of solar-powered steam systems.

6.6 Process and Key Success Factors

> It is the aim of Brahma Kumaris is to highlight the relation between our awareness and the technology we choose. We believe that a holistic approach based on peace, cooperation, and love is the key to a sustainable future (Solar Brochure 2015, p. 4).

The sustainability movement related to the reduction of CO_2 emissions started more than 25 years ago in 1995. It was due to environmental considerations but also due to the general societal and environment context in which the headquarters were placed. It was an environment and society where poverty, scarcity of electricity, shortage of fuel (people felling trees for firewood), and general environmental degradation took place. The climate was tough and plants did not grow easily. Once the trees were cut and the land became barren, it stayed barren. The needs of the growing population were costly to the natural habitat.

A formal department for solar energy was set up in the nineties, and since then, many research and development projects with government support have been carried out. Brahma Kumaris became a registered research centre a few years ago within the Ministry of Science and Technology in India. The research and development initiatives encompass green architecture, photovoltaic power systems, and solar steam cooking systems based on the Scheffler parabolic dish. Several such systems were erected on top of the headquarters to create steam for cooking for the many thousands of participants and volunteers in Madhuban and on Mount Abu. Based on this experience, the research centre has set up INDIA ONE, a 1 MW solar thermal power station, which features 24-h thermal storage. India One was set up by a separate trust, the WRST (World Renewable Spiritual Trust), a daughter organisation of Brahma Kumaris. The WRST has since the mid-90s become a key developer and promoter of steam and energy systems for steam cooking and generation of electricity through PV systems. India One has partly been funded by the Indian and German governments (India One 2019). Today the organisation has installed more than 1400 photovoltaic systems in different BK centres in India. The current aim is to introduce solar energy to all the centres in the country and inspire them to install solar photovoltaic systems for power generation, solar hot water systems, and possibly solar steam cooking systems.

The positive and result-orientated processes encouraged the undertaking of more environmental initiatives and sustainable practices in centres in and outside India. One of the most significant and ongoing initiatives following on from solar power was the Environment Initiative, which has become the umbrella for all their environmental activities. The Brahma Kumaris Environment Initiative was set up in 2009 in Denmark before COP 15. The initiative deals with "awakening greater environmental awareness within the organisation, as well as collaborating and learning from others through dialogue, partnerships, UN conferences, and local initiatives. As a spiritual organisation, the main Brahma Kumaris aim is to help people experience greater well-being through inner peace and higher values" (Brahma Kumaris 2019).

The Environment Initiative is based on five main principles:

- Living with simplicity.
- Buying compassionately.
- Using economically.
- Learning continuously.
- Sharing generously.

The environment initiative grew and spread to most centres around the world. From this initiative, many further activities followed, mainly initiated by the three sustainability leaders but also by volunteers worldwide. Some are described in the following pages.

Local Centres and Network No official environmental department or section was set up in Brahma Kumaris headquarters to deal with sustainability issues, mainly because an official department would require too much administration and manpower. People involved in the initiatives preferred to spend their time developing and implementing them rather than spending time on administration. The sustainability work ,therefore, evolved organically and naturally, so that individuals in the local centres could promote ideas and start activities. The aim was for sustainability to become an integral part of all the departments and centres.

A "green" team or network was set up to participate in environmental activities and to spread information and knowledge. The team inspired other centres to engage in environmental activities whilst allowing for the fact that these were at different stages regarding development and implementation. Currently, more national "green" teams are being set up, and the coordinators can meet once a year to inspire and inform each other to implement sustainability practices progressively. At these meetings, they are updated on developments and bring back information to inspire and empower the local centres. This is supported by current and up-to-date reporting from the sustainability conferences that Brahma Kumaris participates in. More and more groups are set up to work together to create programmes and activities to further sustainability. Senior teachers help to communicate and promote the initiatives. Many seminars, talks, and events are generated to spread the word, both in and outside the centres. As an example, the centre in Copenhagen initiated a yearly outdoor public mediation on sustainability and peace.

Shoestring Budgets The environmental initiatives are based on voluntary contributions (working hours) by individuals, mainly Brahma Kumaris students. Events are organised on shoestring budgets, with contributions from both the local centres and headquarters. The budgets are low, but the output is high. The voluntary contribution factor is an inherent part of the essence of Brahma Kumaris and how the organisation is operated.

Partnerships and Alliances Brahma Kumaris has been working in partnerships with environment organisations for many years. Being part of the COP conferences encouraged collaboration in order to organise side events. Today Brahma Kumaris is partnering with many religious and non-religious organisations. Within the UNFCCC there is an understanding that religion is important when dealing with climate change, and there is now an interreligious panel called the Interfaith Committee that Brahma Kumaris has helped to set up. Representatives from all religions and spiritual traditions gather here to discuss climate issues and form alliances to develop ideas regarding how best to promote and engage in sustainability initiatives.

Other sustainability initiatives undertaken over the years include the following:

Food Consumption Brahma Kumaris participants were and are vegetarians, which is considered essential in implementing sustainability. There is a growing vegan movement within the organisation. Food and ingredients are blessed before being prepared and consumed. This is understood to aid digestion and adds to the spiritual, purifying, and respectful dimensions of daily life and way of being.

Waste Reduction Recently Brahma Kumaris decided to make the organisation plastic free. Plastic bags were banned some years ago by the local municipality on Mount Abu in India. Instead of plastic bags, shopping is carried in cloth bags that can be reused, recycled, and last longer. Additionally, food waste is kept to a minimum. The participants in Brahma Kumaris activities are asked not to put more food on their plates than they can eat to minimize waste. Simplicity, in general, is seen as an essential ingredient in sustainable behaviour.

Yogic Farming Yogic farming has been practised for many years in Brahma Kumaris. Meditation practices are used in food production in the agricultural section of the organisation. Research by SD Agricultural University (Gujarat) indicates that crops which are meditated on regularly contain more nutrients, grow bigger and better, and need fewer fertilisers than crops grown without meditation practices. Crop yields and mineral content have been found to increase by a significant percentage with no extra cost to the farmer or the environment. Other effects that have been observed include an increased sense of well-being and emotional resilience in the farmers. The method is now being practised in other areas of India and is undergoing further research by universities in the country.

Local Radio Station Brahma Kumaris operates a local radio station called Radio Madhuban. Many topics are broadcast for the benefit of the local communities, such as environmental protection, renewable energy, spirituality as well as social initiatives. Twenty volunteers are working in the radio station, which broadcasts 24/7. In addition, Brahma Kumaris has two TV channels that cover all of India. Programmes include spiritual knowledge and practical tips on non-violence and drug-free lifestyles.

> Sustainability is at the heart of everything we do at Radio Madhuban. We aim to make the local community aware of global challenges like climate change, resource depletion, and food scarcity and endeavour to encourage the community to make partnerships to implement innovative solutions to these challenges (Radio Madhuban 2020).

Key Success Factors

The transition toward sustainability was both a top-down and a bottom-up process with an emphasis on grassroots activities. Green became a buzzword. It became the topic that people discussed at the ground level, and that created the impetus for Brahma Kumaris to move the sustainability agenda forward. It was important in the organisation that the sustainability movement grew organically with no power struggles.

Successful Activities

Many activities worked very well. Some of these included talks, lectures, and seminars both externally and internally to gain and promote information and knowledge. Brahma Kumaris attended many environment-related conferences and liked to stay abreast with the latest information and developments in the context of climate change and sustainable development issues. The local centres received visitors from other centres, joined in conferences, and discussed the current issues. The social aspect was important, and the local hosts made sure that guests from other parts of the world were taken on excursions to see interesting sights. This is still so today. The environmental agenda was promoted both locally and internationally. Participation in environmental conferences and events around the world, therefore, became more than attendance at a conference. Instead, it became an event for all involved, promoted inclusivity, and helped further the interest in the environmental agenda across the organisation.

Meditation Green retreats, including meditation and education, were successful, as were the local workshops, where the issues were explained and made understandable. An effective tool for promoting sustainability through personal focus and behaviour was meditation practices. The practice was simple, using easily understood techniques, simple knowledge, universal concepts, and shared values. Values including love, acceptance, absence of anger, and blame, were experienced to be more easily assimilated and adopted through a regular practice of meditation.

Theory versus Practical Implementation Forcing an activity on people or a centre did not work. It had to be a voluntary engagement. The other aspect prevalent in the implementation process, was that some members of the organisation were very hands-on, and some were more theoretical. This meant that one part of the community was interested in tools, skills, and practical application initiatives. The other part was interested in visions of a new world and a new paradigm. Sometimes this gap was difficult to bridge, as each segment could be put off by the other.

Documentation and Research When external funding for research projects was received, detailed documentation of the process, methodology, and financial costs were produced that secured a successful outcome. Detailed project applications, project reports, validity, calculations, and final results were needed when collaborating with governments. Brahma Kumaris became respected at government level due to the meticulous reporting and successful research results.

Long-Term Financial Gain Brahma Kumaris was interested in generating long-term financial benefit, as the institution depended on donations and therefore needed to look at costs continuously. The aim was to generate income from the production of energy in a more efficient and economical way. When a solar hot water system was installed the financial aspect had to be addressed, as well as the environmental and social impacts. To be successful, a reasonable payback had to be secured.

Positive Attitude Having a positive attitude was important and created resilience and strength in overcoming cutbacks, and setbacks and embracing the various practical problems that occurred during the activities. Brahma Kumaris promoted an attitude that included having the right approach, not running away or becoming too frustrated, and not being too pushy or impatient. Go with the flow, it was said. If it does not happen today, it will happen tomorrow, and if it does not happen tomorrow, perhaps it is not meant to happen at all.

The Personal Factor The aspect of "what's-in-it-for-me", had to be addressed. People had to have a good feeling regarding developing better health, having a clear conscience, and generating more happiness from the work they did. For example, "if the discussions are purely intellectual nothing gets initiated and nothing will happen". Regarding this particular aspect, meditation was very helpful in achieving the sense of the world being interlinked, that one was part of a bigger whole, and that each individual had a role to play and a responsibility to the planet. That each individual mattered.

Collaboration and Partnerships Collaboration was a key factor. Positive cooperation and collaboration with partners, both internally and externally facilitated engagement in bigger projects as the collaborative efforts created a strong impetus.

UN Representation UN representation had a profound influence on promoting sustainable development in Brahma Kumaris. The participation in panel debates, interviews, the staging of exhibitions, and discussions with high-level people, including politicians, had a positive impact on Brahma Kumaris, both as an organisation as well as on the people engaging in the activities. One result was that the participants no longer "lifted their eyebrows" when a member from Brahma Kumaris (mostly wearing white gowns) stood up and spoke about soft values and spirituality.

Internal and External Funding The problem areas regarding sustainable development included financial issues such as the costs of covering the development of new projects. The project leader of the solar initiatives pushed for solar energy for many years. When he was able to secure external funding from India and from abroad, the solar power projects were more readily welcomed internally, and became easier to implement.

6.7 Difficulties and Barriers

> Should we buy a new solar system, should it (the funds) go to the global hospital, or should we renovate the kitchen?

The process experienced difficulties where there was resistance regarding the mindset of people within the organisation. Some people understood environmental issues and found these important. Others had difficulty or stated that according to the Brahma Kumaris philosophy, a natural cycle occurred in human evolution and that as a transition was occurring anyway, why engage in environmental issues? More and more, however, the seriousness of climate change was understood, and support for the initiatives increased.

Generational Issue There was a generational issue as people attached to Brahma Kumaris spanned three generations. The Brahma Kumaris senior management team belonged to one generation. At the same time, there were many young people in the centres around the world who belonged to the new generation and had grown up with the problems and effects of climate change.

Limited Funds Brahma Kumaris experienced a shortage of funds from time to time. The question was then how the funds available were utilised, and this raised the discussion of priorities. Which goals should be met first? The discussion was often not only about money but also about the priority of key goals as well as available manpower. It was about spiritual focus and teaching versus practical activities.

Growing Organization Brahma Kumaris was and is a growing organisation. It started as a family undertaking with the elder sisters at the helm and a flat hierarchy. Today Brahma Kumaris has become an international network of organisations, that has to embrace a variety of legal, financial, and administrative issues. Issues such as international law, setting up charitable trusts, rules, and regulations, as well as employees' health and safety regulations, especially in the retreat centres, had to be addressed and dealt with. These considerations and issues absorbed both the time and capacity of the people within the organisation.

Time and Skills As a follow on from the above time became an issue, as people within the different centres sometimes felt there was not enough capacity to implement extra activities and initiatives. The voluntary nature of the work caused a lack of manpower from time to time. This factor was enhanced by the increasing administrative workload as the organisation grew and more and different skills were needed. It created pressure on the existing infrastructure and resistance to engaging in new activities. Essentially, sustainability fitted well within the framework of the philosophy, principles, and values of Brahma Kumaris. There was therefore positive support for sustainability projects from the organisation, and it became understood and accepted that elements including environmental protection and renewable energy were integral to the wider spiritual development agenda.

Reflections

The Brahma Kumaris case is markedly different from the other two examples in terms of content, framework, structure, and financial make-up. It was selected as a case, where both social and spiritual elements were an integral part of the activities of the organisation. The overriding belief system including values, daily practices, and behavioural disciplines is very different from those of mainstream society. It symbolises in its inherent approach to life, the way intentional utopian communities and smaller societies in a similar way have addressed sustainability issues as an inherent factor of their essence and emergence.

Sustainability has, in many mainstream organisations, businesses, and communities, been regarded as an add-on element, interference or a demand to do something different from the so-called norm. There has been a tendency to separate sustainability from the normal way of life. The shift from unsustainable to sustainable behaviour has therefore seemed difficult and has been understood to involve sacrifices. In Brahma Kumaris, however, the inherent vision, philosophy, values, and way of life of the organisation are aligned with the idea and practice of sustainability. The implementation and work with sustainability is seen as a natural extension and evolutionary extrapolation of the Brahma Kumaris principles. The set of values encompasses the spiritual and holistic aspects of humanity and sustainability. These are aspects that generally seem to be lacking in mainstream initiatives but are aspects that are innately connected to the constructs.

6.8 Case Three: Novozymes

This is a fantastic story.

Novozymes A/S is a global Danish biotech company. Its shares are listed on the NASDAQ OMX Nordic Exchange. The headquarters are in Copenhagen, and the company employs approx. 6,400 people. The company operates on most continents. Its main international operations are conducted in China, India, the UK, the US, Canada, Brazil, and Argentina. Novozymes' revenue was 14.3 billion DKK (€1.88 billion) in 2019. The website stated: "Novozymes is the world leader in biological solutions. Together with customers, partners, and the global community, we improve industrial performance while preserving the planet's resources and helping to build better lives". And further: "Novozymes' enzyme and microbiological technologies enable higher agricultural yields, low-temperature washing, renewable fuel, water, and raw material savings" For example, the textile industry can save 70,000 L of water and one ton of CO_2 per ton knitted fabric by using enzymes instead of traditional chemicals. That equals savings of 20–30 L of water and a reduction of 0.3 kg CO_2 for each T-shirt produced. In 2015, the worldwide application of the company's products helped customers reduce CO_2 emissions by an estimated 60 million tons". The company strategy is *Partnering for Impact* based on Sustainable Development Goal nr. 17 in the belief that the greatest impact can be achieved through partnerships with both private and public actors.

The company stated that in 2018 around 5.6 billion people used products on a weekly basis, where Novozymes had added solutions in the form of microbes or enzymes. The company estimated that their customers saved approximately 87,000,000 tons of CO_2 using Novozymes' products. The current strategy and purpose are based on the 17 SDGs, and the company has chosen six sustainability targets to work towards. The keywords are reach, educate, catalyse, deliver, save and enable. The aims the company wanted to achieve by 2020 were:

- Expanding the customer base and reach 6 billion people with biological solutions.
- Educating 1 million people in topics such as biology, sustainability and the environment.
- "Catalysing" 5 partnerships for change by 2020 as the company believes that only through partnerships can global problems be solved.
- Delivering 10 innovative and transformative new ideas that will impact their customers in positive ways and contribute to reaching sustainability.
- Saving 100 million tons of CO_2 in 2020 through the application of their products, thereby also saving on chemicals and raw materials.
- Making sure that all employees have development opportunities in order to achieve their full potential.

Wash clothes clean in cold water.

Novozymes started in 2000 as a de-merger from Novo Nordisk, a global healthcare corporation well-known for its diabetes medicine. At that time, the enzyme business was not developed fully, and no opportunities were mapped. The general prediction for Novozymes (within the group of companies) was that the new company would barely last three years. However, despite hard times, the company developed a brand stating: "Novozymes, unlocking the magic of nature". It engaged in biology-based research. Scientists discovered micro-organisms that produced the enzymes "that can help wash clothes clean in cold water". A vision was created for how biological solutions through enzymes were going to make a profit for the company at the same time as taking care of the environment and people. A set of values were formulated, including social responsibility, ethical behaviour, environmental issues, and credibility. The ambition was to achieve a turnover of 10 billion DKK (€1.34 billion) by 2010 through sustainability measures.

Sustainability was always part of the new company's DNA, inherited from Novo Nordisk, whose CEO for many years had introduced and engaged the company in sustainable development initiatives. In the 80s, there was an international focus on enzymes and their impact on human health. Novo Nordisk invited international scientists and critics to visit the company and discuss the issues. SustAinability (UK) was one of the collaborators at this time and became a consultant to Novozymes during the start-up years. As a consequence, and due to the criticism, the management decided that openness and transparency should be key values in Novo Nordisk, and the focus on sustainability increased. To begin with, the environmental work was carried out by grassroots-type people and green pioneers. The work was seen as an "add-on" and "hippie"-related. But as Novo Nordisk started to gain international respect for its work with environmental issues, attitudes started to change.

The focus and values were inherited by Novozymes. Novozymes started in 2000–2001 with a small environmental team in collaboration with the communication section, helped by external consultants. The environmental part was still regarded as an add-on until 2006 when a new head of sustainability was employed. The environmental department became business-orientated. It grew in size and is today a key and essential element in Novozymes with employees in Denmark and abroad, dealing with decision-making at the executive level, external relations, and communications.

Novozymes' products are enzyme-based. They reduce process temperatures and the use of *harsh* chemicals and water as well as energy consumption. As stated in the introduction, the company managed to reduce CO_2 emissions by an estimated 60 million tons through its customer base. There were many sustainability benefits in the Novozymes enzyme products, and it is this technology measured through life-cycle analysis (LCA) that created the foundation on which the company philosophy, purpose, and activities rest today. Novozymes believes that sustainability must be the purpose for as well as guide business actions.

Vision and Mission

> Sustainability was front and centre of our strategy in 2009.

Novozymes chose to focus on ambition, purpose, and strategy rather than giving a vision or a mission statement and has since the start developed a number of ambitions communicated through purpose statements, goals, and strategies. These have dealt with choosing business directions that embrace both profit, environment, and people. The first was instigated by the CEO and was, as mentioned above: "Novozymes, unlocking the magic of nature" with the goals of generating a profit and doing good regarding the environment and people. In 2009, "sustainability became front and centre" of the strategy, which was quite revolutionary in business terms then. The purpose was to "make the world a better place". Later, in 2015 supported by the CEO and President, a new leap was taken, when sustainability was no longer an add-on but became the reason for being. It became the purpose of the company to innovate, develop, and deliver sustainable solutions globally. No longer only supplying enzymes but also becoming a trusted advisor to customers, providing insight, knowledge, and facilitation. The purpose was to partner for impact and strive to create the "highest positive impact" globally. Novozymes set nine new 2020 targets, where six were related to sustainability, and three were traditional business targets. The belief was that when the company was able to deliver on the six sustainability targets, the financial results would follow. Very unique for a company at that time.

The purpose statement today is *Partnering for Impact,* where the company works to achieve its goals within the parameters of the SDGs in collaboration with governments, universities, NGOs, and customers. Novozymes believes that collaboration is a must if the problems the world is facing are to be solved. Novozymes was one of the first multinationals to take on the 17 SDGs as a strategy for business development, and aims to gradually, over the next few years, deliver innovations that address the needs of the world within the context of the goals. Through the sustainability focus, development, and holistic thinking while also making a profit, Novozymes has become well-known in the world of business. The company was invited to become involved in creating the 17 SDGs by the UN in 2012. The Head of Global Sustainability, who drove the company's sustainability transition, took part in the work on the SDGs and was in 2016 named SDG Pioneer together with eight other people from around the world, chosen among 600 nominations submitted by 100 countries.

Philosophy and Values

> We have been honest and haven't tried to trick anybody.

Novozymes had a strong set of values right from the start of its operations in 2000. It included transparency, trust, openness, and honesty. In addition, there was an adherence to quality of work, attention to detail as well as bringing added value to the market. The company had, through experience, recognised the value of openness and transparency. Where other companies were ruled by mistrust, keeping activities and problems hidden, Novozymes saw how crucial a transparent and scientific approach was to its operation. This openness helped the company ensure that its operations could stand scrutiny by the outside environment and thus contributed to its success. From a practical, operational point of view, the company took pride in operating with a high degree of professionalism and systematic attention to detail. This aspect was also part of the communication and PR activities, which included high-quality of design. In addition, a collaborative approach was taken both externally, in relation to customers and stakeholders, as well as within the company between the departments. The applied set of values and principles contributed to its business success. Credibility became a value associated with Novozymes and its sustainability approach.

Sustainability

> Sustainability is everything we do, not what we say.

In Novozymes the understanding of sustainability was based on the assumption that sustainable development could "save the world" at the same time as "do good business". The company aligned with the Triple Bottom Line model (Sect. 5.2), which advocated the combination of being socially and environmentally responsible and being financially successful. There was merit in sustainability generating money, as this prevented the sustainability discussions from being *hippie-like* to being business-orientated. Sustainability was seen as a serious business factor and became a driver for Novozymes in the context of expanding the market, gaining customers, growing the business, and becoming a market leader. The company's role was also to do what was best for society, as the approach was that sustainability could create "a global society for the long run". The company parameters were therefore wide and embraced issues such as human rights, labour conditions, climate change, and resource mitigation.

Right from the start, the Novo Group of companies wanted to act responsibly and be fair and pay their dues to society, mainly due to the visions of the CEOs. The essence of Novozymes' sustainability philosophy can best be expressed through the following statement from the interview with the Head of Global Sustainability: "My philosophy is that being a responsible company, behaving well, rewarding people, taking care of the environment is a must, is fundamental, is the baseline. But it is not going to save the world. If you want to make the world a better place, you need to innovate and create better solutions that deliver value to people with much less cost to the environment. This is what we target with our new strategy and purpose. We target problems in the world. We find solutions, we innovate and deliver them. We make money and invest more in doing that again and again. Because, it is unlimited. You can give donations and do charity, but that is limited. But if you can make business the driver, the force for good, then it is unlimited".

6.9 Process and Key Success Factors

> How can you make all these things work together, so you can use sustainability to become more successful as a company because you deliver and prioritise innovations for a better world?

The process to change from a traditional company to a company, where sustainability became an integrated factor and a driver for success, both in terms of social, environmental as well as economic perspectives, took approximately ten years. The legacy from the mother company, Novo Nordisk, brought with it a type of product, a set of values, and a way of thinking that paved the way for further development of sustainability thinking. However, without inspirational leaders with drive and ambition, Novozymes might not be where it is today. The process of getting there was a strategic journey and was not taught by outside experts. It was a process that emerged from within the company with valuable lessons learnt progressively. The company went from working with sustainability as a pioneering add-on and environmental reporting, to including it at the heart of all business operations. A strong business approach to sustainability was developed as the driver of innovation, combined with the use of LCA tools to assess and measure the environmental impact of the product range, as well as supply chain products. This approach gave the company the scientific, data- and impact-based and innovative platform from which it could operate and expand. The sustainability profile became a sales- and marketing parameter. The transition process at Novozymes contained several key factors, which are outlined below:

Visionary CEOs and Directors Several CEOs and Presidents of the Novo Group of companies saw the importance and value of pursuing and integrating sustainable development in their businesses, starting as early as the 80s. They gave the head of sustainability the support and freedom to achieve the required results, as top-down support was essential when wanting to implement and drive a company transformation to this extent. The head of sustainability managed to transform the company and embed the implementation process in every department of the business nationally and internationally.

Sustainability Board A sustainability board was set up in 2006, consisting of the vice-presidents representing the key functions of the company covering sourcing, production, sales, marketing, investor relations, communications, and human resources. As they were heads of the departments, there was "a short distance" from a decision being made to it being implemented. When an agreement was reached to implement a sustainable factor, it was the responsibility of the vice-presidents to make sure it was carried out. The head of sustainability proposed ideas to the board, where some were rejected, and others accepted, especially those with an innovative business approach. The board became a strong and essential tool in the transition of Novozymes. The head of sustainability had an opportunity to "discuss priorities based on trends and issues to define strategies, projects, initiatives and to drive the implementation" into operations across the departments. Today, the sustainability board no longer exists as the concept has become an integrated part of all company activities. However, it was a key factor in achieving the success of the undertakings.

LCA: Life Cycle Assessments

> I saw life cycle assessments as the way to understand new business opportunities and expand the scope.

Life cycle assessments (LCA) were first introduced in 2004 in Novozymes to evaluate the environmental impacts of the bio-solution products by looking at issues such as raw material extraction, production, usage, transportation, and final disposal or recycling. The process then expanded into evaluating products from customers and competitors to compare benefits and setting the sustainability criteria. In this way, Novozymes was able to help their customers develop and evaluate their sustainability claims, thereby creating a value-added service for the market. The LCA methodology, also called the cradle-to-grave assessment, became the key lever to translating sustainability from being an add-on to becoming the foundation for the development of Novozymes into a sustainable company.

The environmental assessment methodology meant that Novozymes could claim that their products not only made the customers' products better regarding performance but also saved costs, resources, materials, and energy, thereby lowering the environmental impact for their customers as well as for themselves. This aspect brought the added advantage that Novozymes became both a supplier as well as an advisor to their customers, giving lectures and workshops on how to lessen the environmental impact and integrate sustainability across the board.

The LCA data was gained through own and customers' research, which was then peer-reviewed externally in accordance with ISO14040/44 (Pre-sustainability 2020). The work on and with LCA made Novozymes an expert in the field ahead of its competitors. LCA was applied in a holistic context meaning that all aspects of the products were analysed. The findings were then documented, systematized, and stored in databases.

> We were just in front of the wave……. timing is key. We are in front of the SDG wave. I am a surfer at heart. Global waves are cool…

Timing One particular initiative created an important breakthrough early in the transition, and aided the understanding of sustainability as a factor that could both drive business and gain new customers. It was the initiative by the head of sustainability to approach Walmart, a large international retail corporation. Walmart decided in 2005 to become a sustainable retailer. The sustainability factors gained by using enzyme technology interested the corporation. They saw the benefits and invited their customers and suppliers to hear about the Novozymes products at a conference in 2008. This was the start of a successful liaison, where Novozymes could promote sustainable products to suppliers around the world, validated by Walmart, whilst gaining new customers.

As the quote above indicates, the timing was important. Walmart was approached at the right moment when the corporation was poised to implement its vision to engage in sustainability, but didn't know how to go about realizing the goals. Novozymes was invited to provide both insights and solutions and was at that time ahead of the competition. This gave them an advantage in the market. They were asked to help develop and deliver the 17 SDGs on the global scene right at the beginning. The positive feedback from the PR department encouraged and helped the company further the process of implementing sustainability within its operations as well as gaining new customers.

PR and Communication Good PR and clear and open communication were other key factors in the process. Right from the start, there was an understanding in the company of how important well-designed, open, and clear communication was to the market and in stakeholder relations. The reputation as being one of the first corporations working with sustainability as a business strategy gained Novozymes worldwide attention. The company was one of the actors asked by the UN to help create the 17 Sustainable Development Goals from 2012 onwards and was subsequently asked to present its work at DAVOS and at other international congresses including the COPs.

Clear and Consistent Values From the beginning, Novozymes saw its role as being a successful, science-based, and responsible business as well as a contributor to sustainable development nationally and internationally. A company that was fair and paying back to society. This attitude was communicated to society and stakeholders through its behaviour, communication, and reporting. These aspects had a positive impact on the working environment, where the employees saw their jobs as being meaningful and felt proud of working there. Additionally, there were other benefits such as good working conditions, fair pay, regular health checks, fair treatment, and "good food in the cafeteria".

> It is hard, but the way forward is to continuously recruit people who are smarter than yourself.

Quality of People Sustainability has, until recently and in the main, been driven by passionate and engaged leaders and pioneers. This has been no exception in Novozymes. As mentioned earlier, the CEOs of Novozymes believed in sustainable development and in running a responsible company. They employed the right people to drive the change. The high quality of Novozymes sustainability directors was essential. They were given the freedom and support to drive the change as well as given direct access to the decisions-makers, thereby ensuring that agreed strategies were acted upon immediately and integrated throughout the company. The challenge facing the head of sustainability meant that he had to have a coherent and synergetic team. The team had to know the essence of the business, the working and documentation of LCA, have the right approach to change the mindsets of the different sectors within the company, and have the skills to communicate ideas and initiatives at the CEO level and to customers, suppliers, and society at large. A good team spirit was essential, and the head of sustainability advocated a flat structure, where honest talk was encouraged.

International Profile An additional factor in the positive results of the transition was Novozymes international profile. The transition process placed the company on the world stage. It participated in UN negotiations and was given an advisory role in many big multinational companies. This position, as well as the fame of the people involved, brought the company positive PR and a high-profile reputation. Opportunities to inspire and be the role model for other companies followed in the wake, alongside many possibilities to promote sustainable development through practical examples and case stories.

Other Factors

> If you interview me in 5 years' time, I will hopefully say that the greatest move I ever made was to make Novozymes the first and strongest mover in the company business in relation to the UN SDGs.

Novozymes is now translating the SDGs into areas it can work with, respond to, and deliver on by building systemic methodologies and management tools to make it a reality. The company is gradually, over the next 5–10 years, going to develop and apply innovations to meet the needs of the world as defined and formulated in the Sustainable Development Goals. The company's purpose and strategy are to innovate, develop and deliver sustainable solutions to the world, and to do this in partnerships with public and private organisations and businesses. A new set of long-term sustainability targets will be launched that include social, environmental, and financial goals.

6.10 Difficulties and Barriers

The transition process was not easy as there were no blueprints or other examples of success to refer to. Many barriers and obstacles had to be overcome. Some of the main barriers and obstacles to the process are discussed below.

Scepticism At the beginning, there was a great deal of scepticism toward sustainability within the company. The sales department maintained that the customers did not want sustainable solutions. The belief was that costs would increase, and there would be no added value. There was confusion in the marketplace as to the merit of sustainable solutions. This was combined with a high degree of ignorance regarding the essence of sustainability, and what sustainable development means. A mindset shift had to take place to make executives, employees, suppliers, and customers see opportunities instead of barriers. The introduction of LCA helped to overcome this. By talking business language, presenting data and facts, and proving that sustainability could pay changed the attitudes.

Business "Games" Playing straight, not being arrogant, and not getting involved in the usual business games "probably" slowed the process especially at the executive level. The working atmosphere was in some departments ruled by traditional competitiveness with infights and some degree of backstabbing. Feeling too passionate about sustainability could be a hindrance to discussions and cause resistance and scepticism at board meetings. A certain level of detachment, a hard- core business approach, and persistence had to be applied to gain respect and collaboration from executive peers.

Lack of Resources Realising the goals to implement the life cycle analysis in the company's range of products required new resources in terms of finances. When introducing LCA to the various departments, there were many discussions about finances and resources. Budgets did not initially include extra money for analysis and assessments, and the departments expected the work to be paid for by the sustainability department. When the value of sustainability was established, communicated, and demonstrated, the departments became more willing to collaborate and channel resources for the analysis of their products.

Secret Club In the beginning, the sustainability department was by some regarded as a *closed* or "secret club". The team was small with new people. They were viewed with scepticism, as they were speaking about 'sustainability' and life cycle assessments, which most employees, in general, did not know much about. When the sustainability team communicated their knowledge of business and willingness to collaborate, the mindset changed. As the work developed, more collaboration between the sustainability team and their other colleagues took place, and today sustainability and LCA have become part of the business culture.

Reflections

This is a fantastic story.

From the start, the sustainability transition was not a tried and tested undertaking. There were no KPIs (Key Performance Indicators), no pre-set structures or guidelines on the integration of the construct across the company, although a TBL (Triple Bottom Line) approach was followed. It was, therefore, an emergent process, learning and developing by doing. The sustainability team had to be flexible, convincing, trustworthy, professional, and data-orientated. Respect was gained from the other departments through the presentation of scientific and commercial facts rather than soft value promotion. New systems and structures had to be invented and implemented along the way. The move from employing environmental engineers to hard-core business consultants with a McKinsey background, as well as people with data- and digital expertise, indicated how the development process unfolded. In addition, communication and documentation know-how remained an essential element to talk about the benefits of the enzyme technology and sustainability to both customers, organisations, governmental and intergovernmental organisations, and the world at large.

The quote at the beginning of the report, repeated above, summarizes the way Novozymes transition to sustainability was viewed. A transition that started with visionary and dedicated leaders, supported by a commercially viable, scientific as well as practical methodology through an LCA process. Clear purpose, strategy, and goals were set by the executive team. A sustainability department was set up that had the innovative and psychological acumen to change the mindsets in the company. It evolved and mainstreamed the product range in a sustainability context, introducing marketing and communication parameters that were necessary to drive the change.

The Novozymes story is a success story that is seen as a model for businesses worldwide to be inspired from and follow. It does not guarantee that the company will succeed in the long run from a business/profit point of view. Novozymes shares have had ups and downs on par with other businesses on the stock market. However, the company has achieved a solid, sustainable development platform, for which it is acknowledged. The lessons learnt from the transition process can be transferred and adapted by other companies wishing to engage in sustainable development. Novozymes role as a guide for this has already been established.

The interesting aspects regarding the three cases in this chapter are that they are practical, down-to-earth, and people-orientated. It is relatively easy to write a plan or a proposal. But to carry it out in real life with people who generally have many and different agendas is challenging. We are reminded so many times throughout these stories, of the difference between theory and practice and what matters to people. Many of the people involved in the transition processes were not academics, researchers, or writers but Mr and Mrs on the farm or in the shop trying to make a living while helping to change something. The cases provide an important insight into what matters: for the islanders, the Brahma Kumaris participants and students, and the people employed in the business. It is people, such as these who will make the transition happen.

Not the politicians, or the economists, or the researchers, or the scientists but mainstream people. They need to be heard, take part, be consulted and be looked after.

Readings and References

About Brahma Kumaris (2015–2022) Brahma Kumaris. http://www.un.brahmakumaris.org/. Accessed 2015–2022

About Energy Academy (2015–2022) Samsø Energy Academy. http://www.energiakademiet.dk/en/. Accessed 2015–2022

About Novozymes (2020) Novozymes. https://www.novozymes.com/en. Accessed 2015–2022

Blincoe KS (2020) Essential sustainability: from utopian theory to practical application. Dissertation, University of Kingston. https://eprints.kingston.ac.uk/id/eprint/51146. Accessed 2020–2022

Brundtland Report, Our Common Future (1987) UN. https://sustainabledevelopment.un.org/content/documents/5987our-common-future.pdf. Accessed 10 Jan 2021
Chander LJ (1981) Adi Dev: the first man. Om Shanti Printing Press, Mount Abu
Dawson J (2006) Ecovillages: new frontiers for sustainability. Green Books, Totnes
Environment Initiative (2019) Brahma Kumaris. http://www.environment.brahmakumaris.org. Accessed 15 Nov 2019
De Geus M (1999) Ecological utopias: envisioning the sustainable society. International Books, Utrecht
Garforth L (2018) Green utopias: environmental hope before and after nature. Polity Press, Cambridge UK
GEN (2022) Global ecovillage network. https://ecovillage.org. Accessed 20 April 2022
Girardet H (1999) Creating sustainable cities. Green Books, Totnes
Global Hospital and Research Centre (2022). http://www.ghrc-mum.org/
ISO14040/44 (2020) Pre-sustainability. https://pre-sustainability.com/legacy/download/Life-Cycle-Based-Sustainability-Standards-Guidelines.pdf. Accessed 2 April 2020
Kibert J et al (2012) Working toward sustainability: ethical decision making in a technological world. Wiley, Hoboken
LCA Methodology (2019) Novozymes. https://www.novozymes.com/en/about-us/sustainability/lca/. Accessed 10 Dec 2019
Limits to Growth (2019) Club of Rome. https://www.clubofrome.org/news/publication-of-the-limits-to-growth/. Accessed 10 Oct 2019
Paris Climate Agreement (2022) UNFCCC. https://unfccc.int/process-and-meetings/the-paris-agreement/the-paris-agreement
Radio Madhuban (2020) Brahma Kumaris. https://www.radiomadhuban.in/. Accessed 14 March 2020
Solar Brochure (2015) Brahma Kumaris. https://solar.brahmakumaris.com/downloads/. Accessed 4 Jan 2019
Sustainable Development Goals (2019) UN. http://www.un.org/sustainabledevelopment/sustainable-development-goals/. Accessed 3 Aug 2015
Yogic Agriculture (2019) Brahma Kumaris. http://environment.brahmakumaris.org/component/content/category/40-yogic-agriculture. Accessed 15 Nov 2019

Chapter 7
Barriers Hindering Transition to Sustainability

Abstract The chapter expands the discussion of the realisation of sustainability. The first part addresses barriers to sustainable development that are innate in our current societies, covering political, economic, technological as well as cultural aspects. The second part discusses the barriers that emerged from the case studies in the previous chapter. Some are clearly defined and visible Others are subtle, indirect, and complex. An observation from my research was that the closer I examined the sustainability construct in the context of the existing societal framework the more I uncovered barriers against its realisation. The barriers appear to be as interlinked and as interdependent as the sustainability construct itself. Some are broad, difficult obstacles, and some are practical and basic, easy to understand and address. The difficulty closing the gap between theory and the practical application does not seem to be due to a lack of scientific research or technological inventions and know-how, but due to the magnitude and amount of resistance, found everywhere in our society. Some of the barriers are universal and encompass aspects of our cultural and traditional make-up. Some are specific, such as those experienced and recorded in the case stories. The chapter discusses each group of barriers as well as views on overcoming them.

7.1 The Barrier Issue

> The closer the sustainability construct was examined in the context of the existing paradigm the more barriers I uncovered that hinder its realisation.

In the investigation into sustainability, I examined the work of researchers and writers who have worked for many years on the predicament of sustainability and the aspects that make up the construct. They have described the current dominant paradigm and the emerging crises, and they have discussed factors related to the necessity for sustainable development. They have advocated many models and methods for applying sustainable development. Their data is derived from past and current events, statistics, ground research, and theories of others. Despite this amassed knowledge that includes methodologies, models, and various application tools, some

K. Blincoe, *Achieving Sustainability: The Ultimate Human Challenge*,
https://doi.org/10.1007/978-3-031-10023-9_7

of which I have documented in this book, there is overwhelming evidence that the needed interventions have not been and are still not being made. Since 1987 countless development initiatives have been undertaken. Governments have passed environmental laws, applied various rules and regulations, businesses have talked about corporate social responsibility, triple bottom line, and circular economy initiatives, and civil society has embarked on numerous pioneering programmes and projects. Yet the wide gap between the theory of sustainability and its practical application in a global and universal context has not been bridged and still exists. The gap between the theory and practical application is not due to a lack of research, technological innovations, or other attempts at bridging it, but due to the extent of existing, innate, social, economic, and cultural barriers.

In the months leading up to COP 26 in the autumn of 2021, the media was saturated with articles and stories about the state of the world, the planet, and the catastrophic trajectory towards global warming beyond 1.5°. They were mainly blaming the world's 197 governments for failing to take appropriate and immediate action to curb the emissions and limit the disastrous effect of climate change. Rightly so. As discussed earlier in this book, it has been known for decades that the political and economic system has been moving in a direction that is undermining the planetary boundaries. The system has undermined the livelihoods of millions and is destroying the biodiversity necessary for the well-being of future generations. Governments have taken initiatives, but too few and too fragmented. They have made no overall master plan to secure the future well-being of the natural habitat. With regard to the recent catastrophe, the pandemic, this lack of foresight could be excused, as the virus was unexpected and spread fast. But global warming, on the other hand, has been predicted to threaten humanity since the fossil fuel corporations in the US became aware of its devastating impact in the late 70s and 80s. They even acknowledged that zero emissions could be reached by 2035, or at the latest by 2060, if action was taken at that time. That precaution was not heeded, as we know (Rich 2019).

However, after the Brundtland Report was made public, and the UN called the nations together in the first COP in 1995, some initiatives were taken to embrace sustainable development. The progress that has been made since 1987 has mainly been through initiatives taken by the UN followed up and supported by a variety of actors such as civil society. Governments have until recently considered the neoliberal political and economic system to be beneficial to society to such a degree that top-down policies and interventions regarding environmental problems have largely been swept aside. Instead, new policies have, in the main, dealt with rewarding the electorate here and now. In the instance of the European countries, new policies have been about protecting industry, supporting agriculture, and promoting well-fare policies. The sustainability agenda has until recently been an add-on discussion.

7.2 Key Barriers to Sustainable Development

> Achieving sustainability in the real world presents a daunting and complex challenge (Kibert et al. 2012).

There are many significant barriers to achieving sustainability, both rational, visible, and subtle, as mentioned above. Many are ingrained in our societies. Some are barriers generated centuries ago, often based on religious beliefs, and others are recent and a result of societal advances where new lifestyles and changing attitudes have impacted on our values and principles. There are also some in the category of universal barriers that encompass aspects of our cultural make-up. Most barriers are structural, and some are procedural. Overcoming them will require both structural change as well as new procedural measures. How a transition is evoked in this context is discussed in Chap. 3 on development, where the traditional economic models are compared with the model of sustainable development. Opportunities regarding overcoming barriers are addressed in the chapter on cases (see Chap. 6). The feedback from their processes adds a valuable local perspective to the barrier discussion, and it also points to groupings, such as the universal or global versus the individual or local. Both levels are critical to the implementation of sustainable development, and both enhance the understanding of what is blocking the transition process as well as overcoming the obstacles.

The barriers are interlinked. Discovering one leads to the next, in the same way as overcoming one barrier might lead to overcoming another. To illustrate and simplify the barrier discussion, I have included two lists. One demonstrates the summary of universal barriers, that I came across in my thesis research, and the other is a summary of the barriers discovered in the case stories. They are not revolutionary discoveries, but the fact that they are ordinary is interesting, as they serve to point out and remind us of that which we must deal with in everyday life and the vital role the ordinary plays in the change game. These barriers also tell us that simple methods when dealing with sustainability are just as important as complex technological solutions. The lists are helpful when wanting to draw up a framework for engaging in sustainable development initiatives and what to be aware of when met with resistance.

7.3 Universal Barriers

My research pointed to the following universal barriers:

- The concept of sustainability—the complexity of the construct.
- Operational models and methods—there are no universally agreed procedures.
- Political barriers—business as usual attitude.
- Economic and business barriers—costs and financing.
- People-related barriers—resistance to change.
- Technological barriers not advanced, practical, or far-reaching enough.

- Lack of knowledge and education—ignorance is a barrier.
- Lack of positive visions—nothing that inspires.

The Concept

The concept of sustainability is an ideal or even an ideology. It is a fairly easily understood universal approach to the issues at stake. It is defined in the Brundtland Report and advocates an intention and a goal to work for a new direction of development. Different from the traditional development goals coined by Harry Truman in 1949 and subsequent theories (see Sect. 3.1). Both sustainability and sustainable development are in the report described in universal terms. As such they are theoretical—ideals to work towards, promoted through sets of declarations and intentions that were not binding but voluntary. To be realised, the concepts need to be legally binding and translated into practical manuals. Both the voluntary and the theoretical aspects have, therefore, been barriers to implementation as it has been too easy and uncomplicated to agree to the declarations, blame complexity, and do nothing.

Sustainability represents a new radical idea of a society that is different and better than the existing one. It can be compared to and regarded as a utopian ideal. Visions of better futures have always occupied our minds. Many new ideas have emerged over time, and some have been implemented. Experiments have failed, and some have succeeded, as this book documents. The challenge is that sustainability has now become a universal necessity requiring the Race to Zero Emissions, mitigating climate change, reversing biodiversity loss, regenerating local livelihoods, and producing and consuming sustainably. Sustainability is no longer just an idea for a new development theory but is a model for society that must be implemented. Failure to do so is not an option. To work, the constructs must be translated into operational and practical processes for implementation purposes.

The difficulty of translating sustainability into a practical model, comprises several aspects. The UN Brundtland definition is broad and open to interpretation and it encompasses nearly all life. It is universal and non-specific, and consequently, a multitude of interpretations have emerged that leave the concept open for misuse in a variety of contexts and guises. The high degree of complexity involved in generating sustainable practices is also a critical barrier to successful implementation. As a result, implementation attempts have emerged from all directions at many different levels in society, seemingly without coherence. The lack of universally accepted operational models and methods has also prevented successful implementation. The latest model, the 17 Sustainable Developments Goals, has been generated by the UN as an attempt to simplify the task by breaking down the construct into 17 goals.

The goals were designed as a guide or pathway and contained sub goals and indicators as an aid to implementation. The model was introduced in 2015. Since then, many sectors in society have adopted the goals as a framework for engaging actively with sustainable development (see Sects. 2.3 and 3.5) Unfortunately, this has also seemed to be unmanageable. The task for a municipality, for example, to deal with 17 focal points simultaneously is challenging. Often a choice will, therefore, be to focus on goals that are easy to reach—such as introducing sustainability topics in schools—and delaying working with those that are difficult and costly such as expanding the public transportation system. The cost has become an issue and a barrier, either real or perceived. As sustainability covers all aspects of environmental, social, and economic activities, it has been a challenge to decide what to focus on in the absence of a master plan or national governmental framework.

Political Barriers

Many complaints have been publicised on the lack of political will and commitment over the years. The voluntary and non-legal aspects of the agreed UN declarations have made it possible for governments to avoid taking the necessary steps. The legislative initiatives that have been made have not been implemented systemically or coherently but have been limited to relatively small and fragmented initiatives. Some regard it as greenwashing to appease the green lobbyists. Additionally, the developed nations have been asked to carry the main responsibility for helping developing nations pay for their initiatives. This has been lacking even if the rich countries have benefitted the most and are still benefitting from the business-as-usual paradigm.

Numerous promises have been broken over the years. More recently, as an example, countries have reneged on promises such as honouring the Paris Climate Agreement made in France in 2015 (Sect. 2.4). Furthermore, at COP 25, which took place in December in Madrid, 2019, some of the big nations pulled back from agreeing to meet the 17 Sustainable Development Goals within their set time frame. So, at COP 26, in the wake of the dire findings from the latest IPCC 6th Amendment Report, a flurry of re-commitments, new commitments, and pledges were made by governments, financial institutions, and other actors. However, the public is sceptical, and the feedback from the Conference was that this was not enough and that it was too late (Sect. 8.4). People have lost confidence in politicians, and for a good reason. As this book keeps stressing: it is the failure of governments to act that has caused CO_2 emissions to rise so fast and contribute to climate change, chemical pollution, misuse and overuse of resources, and loss of biodiversity. A failure that could have been prevented had the precautionary principle together with corrective action been taken decades ago when the evidence accumulated both scientifically and in real-time. The state is responsible for the well-being not only of the present but also of future generations. The inability or lack of will to make the necessary interventions has been and is the absolute key barrier to sustainability. However, we, the citizens, are also responsible, as I believe our political leaders are a reflection of the people, who elect them. We earn our leaders. We need to elect politicians who are willing and capable of applying the appropriate solutions to address the crises. That is something positive and proactive that we can and must do.

Economic and Business Barriers

The business sector is an integrated part of the current system as it deals with the production and consumption of services and goods supported by investment and transference of wealth. The key goal of corporations is to safeguard their shareholders and generate wealth in monetary terms. The business sector has therefore found sustainable development challenging and burdensome and has frequently maintained that sustainable development is too complicated and too costly to deal with. However, in recent years the business sector has begun to realise, due to market pressure and public opinion, that it needs to change its focus and be seen to take a real (and not feigned) interest in social and environmental issues. Another problematic issue relates to business versus natural resources. Until recently, natural resources have been regarded as free to use and exploit without repercussions. This aspect is inferred previously (Sect. 3.3) where I write about the crises leading to the emergence of sustainability. Natural resources represent one of the vital crisis areas. This laissez-faire attitude to resources still prevails in the industry sector. Resources, especially non-renewables such as minerals and metals constitute a core element in the Brundtland definition of sustainability. They are exploited and not protected as part of the natural capital.

Business is, however, making attempts to protect the non-renewable resource base, specifically through the application of cradle-to-cradle and circular economy models (Sect. 5.2). The progress is slow. Despite good intentions, the effects of these models are elusive, meaning that any positive results do not seem to have impacted the resource base, as yet. The models do not advocate cutting back on consumerism. Instead, they promote production and consumption by promising a high degree of recycling and up-cycling. As there is a limit to the degree of recycling, that a material can withstand, the problem of mining for new resources still exists. The rich countries should reduce consumption as well as encourage recycling. Both are necessary.

The business sector is aware that it must be seen to engage in green and responsible production. It has launched many pioneering initiatives ranging from small and exclusive second-hand clothes shops and other similar business ventures to working on greening big businesses. As described in the third case story (Sect. 6.8) Novozymes is one of the pioneering Danish corporations that has applied sustainable development models and analysis tools. The company has successfully used the Triple Bottom Line model as well as LCA applications in their product range and has proved that it is possible to work towards sustainability, and being a business.

The sector has struggled with the difficulty to systemise and standardise operations including supply chains, as part of their transition initiatives. Generating effective and valid measures and monitoring performances, particularly regarding social initiatives, have proven difficult. My research indicates that measuring social impact has been and still is problematic and provides a dilemma regarding maintaining the focus on the economic bottom line. The argument is that some sustainability measures that relate to moral and ethical principles cannot be quantified, nor can they be compared to the financial results of a company's operations.

To reach the sustainable development goals top-down as well as bottom-up implementation efforts must be symbiotic and coherent, supported by sufficient financial investment. The Global Financial Markets Association estimated in late 2021 that limiting global warming to 1.5 °C would require an overall investment of US$4–5 trillion annually, as previously stated. The sheer scale of financing and investment has for many years been seen as an insurmountable obstacle to sustainable development, both from a political and from a business perspective. It is in this context worth noting that the pandemic, Covid-19, has induced governments to allocate almost unlimited funds to take radical measures to alleviate a looming economic meltdown. As a comparison, it is also worth remembering the cost of the damage that climate change has already incurred as well as the future cost in money and human terms. The finance to address and alleviate the environmental crises must be found and generated from public and private sources. A new economic model must replace neoliberalism, where the focus changes from economic growth to growth of human potential and repair of the biosphere. The economist, Kate Rahworth's Doughnut Economics model comes close to embracing the needed factors (Sect. 3.6).

A final, slightly different perspective to be mentioned in the business section on barriers is the power of the wealthy elite. Naomi Klein, the pioneering Canadian activist and author, has been addressing aspects of the business sector for many years. In her books such as "No Logo" from 2001, "Fences and Windows" from 2002 as well as the more recent "This Changes Everything" from 2014, Naomi Klein essentially holds corporations and the powerful elites accountable for the mess we are in. She maintains that they have and are benefitting from the unsustainable status quo to a highly destructive degree. The elites consist of powerful corporations, magnates, and other influential economists and business executives, whose aim is to intentionally resist sustainable development due to their single-minded focus on profit, wealth creation, and control without regard for the consequences.

Cultural Barriers

Many critical obstacles to sustainable development are revealed in people's behaviour. This includes the factor of population growth. The world population is predicted to increase to approximately 11 billion by the end of this century, according to the UN. This increase constitutes a challenge to the health of the planet as it adds a further strain on the planetary boundaries. The consequence means increased consumption, resource depletion, CO_2 emissions, and pollution. In addition, there is a predicted increase in the middle-class segment worldwide which will also add to the burden. To overcome the first barrier, strong governmental leadership as well as increased and intensified education opportunities are required, especially for girls and women. The IPAT Equation (Sect. 5.2) proposes that to offset an increase in population and affluence, the unit of resource use per person must decrease dramatically, and technological impact in terms of solutions must increase proportionately. It is difficult to assess how the equation will work in practice, and as such, it is only useful as a rough guide. The real question is whether the natural resource base can carry a population growth as well as an increase in affluence. Resource use should be significantly reduced by the rich countries to allow for population growth and an extended middle class but whether this can happen without political intervention is doubtful. Both predictions represent problems and will aggravate the current crises.

Culture and traditions play a critical role. Michael Harvey (Sect. 4.1) talks in his book, 'Utopia in the Anthropocene' (2019), about a set of values, which he calls the "CIMENT mnemonic". They include competition, individualism, materialism, elitism, nationalism, and technology. The values, Harvey claims, are intrinsic to the pursuit of monetary growth in the current world order. He maintains that there are only winners and losers as "the only choice appears to dominate or to be dominated"

(2019, p. 36). The drive for individualism has spread over the last few decades. It is the essence of the American dream. Individualism is an integrated part of the political and economic ideals introduced by Reagan and Thatcher, inspired by the Russian/American writer and philosopher Ayn Rand (1905–1982), the heroin of neoliberalism, rejecting altruism and promoting egoism. Materialism is viewed as a symptom of the growth paradigm. The phrase *I shop, therefore I am* has often been quoted. Accumulating material wealth has been regarded as more important than protecting the natural resource base, the natural capital. Elitism, Harvey's fourth value has become widespread and accepted, seen through the increase in power and influence in the executive class and by a small select group of industrialists and economists operating in the global arena. It is this dominance and control the author Naomi Klein addresses in her books, as mentioned in the section above. The surge of nationalism that Harvey mentions has recently been felt in most countries expressed through, for example, the blind fear of foreigners, refugees, and immigrants, including an increase in racism. Closing borders, building walls to keep people out, and turning the focus from the cosmopolitan *me* to a national *us* has been one of the consequences of this. As the last factor in the CIMENT set of values, Harvey mentions technology, which I address in the section below. Even though technology is regarded by many as the essential problem-solving tool to fight climate change and other issues, it is also a field that has presented our societies with challenges and obstacles preventing sustainable development. Overcoming the CIMENT values will be difficult. They will change the day the world or society changes, and a new political and economic order and focus will substitute the old. A symptom of this can be found in the impact of the pandemic, where the *I* changed to *we*, and where the virus impacted everyone with the same force, whether part of an elite group or not, whether wealthy or poor. Less individualism and less elitism were positive results of the pandemic.

Apart from Michael Harvey's CIMENT values, there are additional, cultural and psychological factors to be included in the section on cultural barriers. Religions, traditions, and lifestyles vary from country to country, and many contain elements that must be changed or overcome if we want a transition from an unsustainable to a sustainable state. In some religions, such as Christianity, nature has been regarded as an element to be controlled. In others, there is no equality between women and men, where the male dominates and controls the female. Some religions promote eating meat, which is now considered a detriment to sustainable behaviour. Denmark has, as an example, always had a tradition where celebrations and festive seasons have been associated with high consumption of meat, and the current anti-meat campaigns have challenged the population. Festive seasons are associated with increased consumption of consumer goods. Christmas is a classic example of this. Religious festivals such as the Chinese New Year, Deepavali etc., are also consumption orientated, as are most celebrations. During the might of the neoliberal era, from the 1980s until the start of the pandemic in 2019, the consumption of consumer goods rose dramatically, and any celebration became associated with material purchases. Lifestyles became based on materialism related to fashion, household and electronic goods, travel, and entertainment. Shopping became the universal therapy. If you felt good, you bought something. If you felt bad, you bought something. Shopping became a lifeline, and the advertising industry had golden days.

As sustainable development challenges and impacts our cultures, values, traditions, and lifestyles it is bound to create resistance and reluctance, even though people, in general, have an emotional and theoretical understanding of the issues at stake and feel responsible. In this context, it is important to be aware of the power of emotions. To blame ordinary citizens for the climate crisis and accuse them of being irresponsible will not produce changed behaviour. Neither will self-blame nor blaming one's neighbour. Looking back in history, radical change has occurred either through wars, uprisings, rebellion or military coups. In the instance of sustainability smaller intentional communities have created radically different ways of living according to their principles. Many have been successful because they were created by consensus, an agreement where the members of the community had a shared vision and joined forces to realise it. These initiatives have empowered people, and when we feel empowered and included, are heard and respected, and are invited to participate and make a difference, change can happen in very positive, constructive ways and extraordinary ways.

I believe that the pandemic has helped pave the way for a new direction. New initiatives and entrepreneurship are emerging everywhere that are leaving the old patterns behind and starting new ones: in the food industry, where vegetarianism and veganism are on the increase even in countries with strong meat traditions, and I see food waste reduced. In the fashion industry second-hand clothes and recycled textiles are becoming the new trend. In the transportation industry international travel by rail is becoming a trend, and the electric car is replacing the fossil fuel one. I see it in a turn from consumer living to simple living. Having less, buying less. Designers are becoming sustainability orientated, the advertising industry (the biggest nudging factor in consumerism) is, at last, changing its ways, and businesses are discovering new opportunities in the transition movement. This is positive. The psychological factor is incredibly important. If mainstream people get interested and can see themselves in the new scenarios and can see benefits for themselves and their families, massive change can occur which can overcome any barrier.

The Technology Barrier

Technology is frequently mentioned as the factor that can best be utilised to meet sustainability challenges. Hence it is also the lack of technological solutions that are currently regarded as the main barrier dealing with climate change, for example, in terms of reduction of CO_2. Many innovative ideas based on technological innovations have been mentioned in this context, such as CCS (Carbon Capture and Storage), which is a method to extract CO_2 before it enters the atmosphere, turning it into a liquid form and storing it in the underground often in used oil and gas reservoirs. It is a technology that has been applied for some years already. The problem is the amount of CO_2 needs to be stored continuously, and there is a lack of methods to develop products using CO_2. In the car industry, the precious metals used and the limited life of batteries represent challenges that must be resolved, and substitutes found. The challenge to develop operational models for sustainable development rests to a large degree on technological advancement in the primary sectors, such as the transportation, building, production, and agricultural sectors. Most models and methods relating to the implementation of sustainability are based on new inventions. Technology is essential in meeting many challenges both in the short-term and long-term, but the reliance on it to solve all our problems creates bottlenecks. Research takes time and costs money, and the global challenges that need to be addressed and met immediately, cannot rely on nor wait for new inventions. People can be nudged into changing their lifestyles and purchasing habits in a less resource-consuming direction that does not require technology. Recycling can be done with existing methods. Applying a tax on CO_2 emissions does not need the latest technological fix. Many tried and tested older methods can alleviate the current problems. The need for new technologies may, to a certain degree, be artificial, probably used as a delaying tactic by actors who do not have an implementation plan.

Lack of Education and Knowledge

An additional societal and critical component that hinders the green transition is the general lack of knowledge of the field. It is also one of the observations made in the case stories in Chap. 6. Education was, in the case stories considered to be both a key element in the transition successes as well as being an obstacle, where there was a lack of relevant knowledge. An important factor in this context is the lack of education on sustainability in schools and colleges worldwide, which means that people in the main are uneducated on the topic. The language around sustainability and sustainable development has often been jargon. The concepts have not been understood by the mainstream, and the field of sustainability has, as a consequence, been open to misinterpretation and misuse. This situation has opened the door to the creation of myths, conspiracy theories, and fake news. It has led at one extreme to increased resistance to sustainability, and at the other, to a sense of frustration, desperation, and aggression. As the awareness of the topic has grown, so has the need for education and knowledge. People want information, guidance, tools, and skills to play their part in a transition process, as many feel personally responsible for the state of play. Education in sustainability does not exist in most schools, and many people around the world still do not know what sustainability or sustainable development means or entails.

It would be relatively easy to include sustainability in the school curriculum from primary through secondary school to further education. Sustainability is not part of the official curriculum in Denmark, but the situation is being addressed, and it is my hope this will be introduced within a few years. So far, a few NGOs and other organisations, such as Chora 2030, offer modules to primary and secondary schools on the 17 Sustainable Development Goals. The schools are keen to embrace the Goals. The ideal would be for sustainability to become an integrated part of all levels of education.

Lack of Positive Visions

The chapter on utopianism mentions the lack of inspirational scenarios for an alternative and sustainable future. The utopian communities that did get established contained within them clear ideals and positive visions for how a sustainable future might look, creating a common inspirational platform that held and promoted collaborative efforts. Images of positive futures are necessary to inspire people to act. Unfortunately, it seems that people over the past century have found it easier to imagine dystopian rather than utopian ideas of how the future will emerge. Books such as "Climate Wars" by Gwynne Dyer (2011), "The World Until Yesterday" by Jared Diamond (2012), "Six Degrees" by Mark Lynas (2008), "The Ends of the World" by Peter Brannen (2017) and "The Uninhabitable Earth" by David Wallace-Wells (2019), are just a few of the recent apocalyptic scenarios we are presented with. The books present a reality that instils fear and worry. However, the truth is necessary. Unless we know what we are faced with, we can't act appropriately. The question, however, remains whether the fear initiates a positive and constructive action or has the opposite effect and creates hopelessness and paralysis. I believe that the dystopian scenarios promoted by the entertainment industry can have a negative impact on our psyche and create invisible barriers to bringing about the necessary positive change. We need optimistic, visually inspiring scenarios for the future. Scenarios that we understand, feel inspired by, and can identify with. The small intentional communities I write about in Chap. 4 on utopia had visions that people could identify with, see themselves in, and benefit from. The same evidence is found in the case stories where everyone participating was inspired and engaged because the ideas appealed to them, because they could identify with the achievements and because they could gain from the change.

Many maintain, including me, that humanity is faced with choosing between two different trajectories of the present: A utopian future consisting of a new global, sustainable paradigm or a dystopian future where no change of direction occurs and where humanity is faced with possible extinction through a multitude of disasters. One of the recent symptoms of the latter is outlined in the work of the UK Professor Jem Bendell. He expands on his predictions in his Deep Adaptation work, where he asks his followers to embark on a journey of preparation for the coming derangement which may cause societal collapse within the next few decades. Bendell and his colleague Katie Carr propose four Rs in the Deep Adaptation framework: resilience, relinquishment, restoration, and reconciliation with which to deal with the coming collapse (Bendell 2021). The work is relevant as many young people fear the future, and climate anxiety is spreading rapidly. The fear factor is a reality. No one can predict the future, but some signs point to a potential societal collapse due to the threat to the planetary boundaries and intensifying climate impacts. The key task is to realistically assess what can be done collectively and personally. Plan how to go about it—and then take action.

7.4 Case Story/Local Barriers

The main section of the chapter deals with the broader and more universal barriers and obstacles. The following section contains the significant barriers revealed in the three case stories. They are specific, based on real-life issues, and address mainstream situations. Even though the three cases were very different from each other as they consisted of an island, a faith community, and an international corporation, many of the factors experienced both success factors as well as barriers in the transition processes, were similar. The barriers refer particularly to operational problems such as lack of relevant skills and competences, lack of funding, and other practical aspects. More than half of the problems were people-related, experienced as conservatism, lack of participation, pressure to succeed and negative gossip. Universal elements such as politics, technology etc., did not play a significant role in the local contexts, only as an external influence. The factors in the list below refer to the cases but can be used as a guide for other similar initiatives due to the commonalities that arose in their diverse contexts.

Case-Related Barriers

- Conservatism and scepticism.
- Lack of relevant skills and competences.
- Lack of funding/financial pressure.
- Lack of participation.
- Pressure to succeed.
- Negative gossip and power struggles.
- Being too passionate.
- External factors.

Conservatism and Other Resistances

There was resistance to change in all three cases. For some, it was due to conservative and fixed mindsets. For others, the generation gap between the older and the younger generations was pointed out and caused tension. This was specifically so in the context where the leaders were elderly. In the business context, the sustainability team was regarded as closed, slightly hippie-orientated, and not mainstream. The key tools to overcome this type of resistance were to be inclusive, provide ample communication, show results, and prove that sustainability could make a profit and make a positive difference.

Relevant Skills and Competences

There was a recognition across all the project teams that having people with the right skills and competences was essential. In some instances, it was felt that having what was thought of as the *wrong* people was seen as a barrier factor, especially regarding being positively engaged and participatory. The aspect of involving voluntary people in a change project can cause problems. Most organisations with a visionary goal will attract volunteers or helpers. After a while, resentment can arise due to not being paid. The Brahma Kumaris case was different in this respect. Here a conflict occurred from time to time regarding the priority of what to focus on and a lack of the right skills, due to the expanding organisation. Novozymes had the financial ability to employ the right people at the right time, which they saw as an essential factor in achieving success. Being able to bring people on board with new and relevant skills was crucial and was seen to be a vital factor in the ongoing successful development. Getting *the right people on board* can be difficult, especially in a transition process, where a new direction is being charted that has neither been tried nor tested. The key in this situation is to be flexible and make changes as the process unfolds and new competences are required. The training of participants, volunteers, or staff is an important factor in this context.

Inclusivity and Participation

Participation by people at all levels of the undertaking was seen to be essential. Lack of participation was therefore regarded as a barrier to achieving the goals and objectives and could indicate that the process was lacking in something that had to be addressed. To alleviate this, aspects such as inclusivity, continuous and ample communication, and information-sharing were needed. The *what's-in-it-for-me* factor also played a significant role in terms of people's interest in participating as well as feeling valued, appreciated, and respected. The human aspect is often overlooked when embarking on new initiatives. It is generally the leader or management that initiates the change, and employees or participants are the critical levers for implementation. Their active and positive engagement is crucial for successful participation and achievement of the set goals.

Inclusivity and participation are crucial in all collaborative initiatives. Some of the sustainability projects mentioned in Chap. 4 on utopianism, such as Masdar and Songdo, illustrate what can happen when there is a lack of inclusivity and citizens are not participating in the processes. The projects, although beautifully designed and incorporating the latest and most advanced technology, failed to attract the people they were targeted to house. The participatory aspects were left out of the top-down processes that were undertaken. People were not asked or included in the design considerations.

Pressure to Succeed

The three case teams felt pressurized in terms of generating results and succeeding within a set time frame. Sometimes a certain amount of pressure can be constructive to a process. However, too much pressure leads to stress and burn-out and can have a negative impact on results. When a process is emergent, and there is little or no pressure to achieve results within a specific time frame, the people involved are more at ease, but this is a rare occurrence in any project context. There is nearly always pressure to meet the objectives, succeed, or perform. This pressure can be alleviated by progressively increasing the competences and support systems necessary and having the ability to employ the right people.

Negative Gossip and Power Struggles

Negative gossip was pointed out as barriers in the case stories. It was felt that back-stabbing, negative gossip, and internal power games were destructive to the processes. I have led several organisations and have found that there will always be some negative gossip and power games being played out in a team or working groups, whether in small or large organisations. It is necessary to ensure that this is dealt with and contained, as the toxicity caused can be very destructive to an organisation or a group of people. This is specifically damaging when significant and large-scale change is occurring, such as in the case stories. Tools to avoid harmful gossip include clear, frequent, and relevant communication, regular participatory meetings, inclusivity, and shared responsibility. Authenticity and straight talk by leaders help curb power games and power struggles. Clear roles and responsibilities are important. Successes and failures should be shared by the team and not be placed solely on one individual. Having a motto of *we* rather than *I* is constructive.

Too Much Passion

The head of sustainability in Novozymes mentioned too much passion as a potential barrier to success. He felt that his keen passion for sustainability hindered the implementation process and thought that had he been more detached, he might have overcome the obstacles more easily, particularly in dealing with his superiors and peers. I have similarly been too passionate in different contexts of my career. Some passion is constructive, but if you are fixed on one way or one set of principles, or one specific goal without awareness of what your external context is, you will meet with resistance. I, as an example, introduced the 17 Sustainable Development Goals too early as a framework for the organisation, I was leading in Denmark. The surrounding environment, that our organisation depended on for financial support had no clue about the goals, as they had just been launched. They did not understand the framework and therefore withdrew their support. A certain amount of detachment and a laid-back approach is necessary, as is the knowledge of who and what you are dealing with for support—financial or otherwise. It was a hard but important lesson to learn.

External Factors

External factors are outside the control of the people running initiatives, but they can seriously impact the processes. These can be top-down interventions such as new laws, legislations, or international events. The external environment cannot be controlled, and a certain degree of flexibility must be allowed to embrace such interferences. Development projects are often under pressure as discussed above due to a lack of funding and/or financial support. Practical, physical, and visible demonstrations of the merits and results of the project's progress can ease this pressure and can lead to more support both internally as well as externally. The ability to be flexible is essential when dealing with new, emerging processes.

Other Lessons

Additional points included topics such as the importance of the size of the project and size of the group of people involved. Having less influence over external events can present problems as, can problems relating to theoretical versus practical aspects. The island team said that to succeed a certain size of community would work best. They also wanted more local responsibility, more independence and less government interference. Brahma Kumaris pointed out that a gap existed in their organisation between the theoretical versus the practical aspects of sustainable development. The attempt to bridge this gap was experienced as a challenge. In addition, they stressed the importance of meditation as a tool for personal or inner sustainability, which they regarded as an integral part of sustainable development.

7.5 Perspectives

> Sustainability, or the ability to sustain humanity and the biosphere, is humankind's greatest challenge and one that we must face using creativity, imagination, and understanding (ICIS 2002).

What seems to be the most critical elements required to overcome problems when embarking on sustainable development initiatives are, firstly and simply the determination and the will to do so. Barriers can be overcome, and problems can be solved if the motivation is strong enough. The barriers to sustainability that are embedded in the current societal structure are connected and interlinked. The profound systemic transformation that is needed will not happen purely by applying new technologies or new models. Many other factors need to be taken into consideration, especially the human-orientated or psychological factors which can only be applied if a holistic approach is taken.

Examining the barriers thoroughly and assessing the priority of which to overcome, when and with what effect, is necessary. The barriers and obstacles addressed in this chapter can serve as a guide to some of the more obvious resistances that will occur when embarking on making profound changes. Some are more important to overcome or deal with than others depending on the nature of the project, its context, and environment. The teams in the case stories said that they would probably apply a similar approach were they to embark on similar projects. However, many of the oppositions, conflicts, or barriers encountered could have been avoided with some prior knowledge of what might occur.

Readings and References

About Brahma Kumaris (2015–2022) Brahma Kumaris. http://www.un.brahmakumaris.org/. Accessed 2015–2022

About Energy Academy (2015–2022) Samsø Energy Academy. http://www.energiakademiet.dk/en/. Accessed 2015–2022

About Novozymes (2020) Novozymes. https://www.novozymes.com/en. Accessed 2015–2022

Bell S, Morse S (2008) Sustainability indicators: measuring the immesurable, 2nd edn. Earthscan, London

Bendell J (2021) Deep Adaptation. https://jembendell.com. Accessed 5 January 2021

Blewitt J (2008) Understanding sustainable development. Earthscan, London, p 157

Blincoe KS (2020) Essential sustainability: from utopian theory to practical application. Dissertation, Kingston University. https://eprints.kingston.ac.uk/id/eprint/51146. Accessed 2020–2022

Braden G (2014) The turning point: creating resilience in a time of extremes. Hay House UK Ltd., London

Brannen P (2017) The End of the World. Oneworld Publication, London
Brundtland Report, Our Common Future (1987) UN. https://sustainabledevelopment.un.org/content/documents/5987our-common-future.pdf. Accessed 10 Jan 2021
Caradonna JL (2016) Sustainability: a history. Oxford University Press, New York
Carey M (2021) Value(s): building a better world for all. William Collins, Croyden
Circular Economy (2015) Ellen Macarthur Foundation. http://www.ellenmacarthurfoundation.org/circular-economy. Accessed 5 April 2015
Chora 2030 (2022). https://chora2030.dk/. Accessed 28 April 2022
CSR (2015) CSR. http://csr.dk/de-gode-penge-ligger-i-b%C3%A6redygtighed. Accessed 15 Sept 2015
Diamond J (2011) Collapse: how societies choose to fail or succeed, 2nd edn. Viking Penguin, New York
Diamond J (2012) The world until yesterday. Allen Lane, London
Dyer G (2011) Climate wars: the fight for survival as the world overheats, 2nd edn. Oneworld Publications, Oxford
EEA 5-year Report (2018) European Environment Agency. https://www.eea.europa.eu/publications/environmental-indicator-report-2018. Accessed Jan 2019
Ehrlich PR, Holdren JP (1971) Impact of population growth. Sci New Ser 171(3977):1212–1217
Elkington J (1999) Cannibals with forks: the triple bottom line of 21st century business. Wiley, Hoboken
Farley HM, Smith ZA (2014) Sustainability: if it's everything, is it nothing? Routledge, Abingdon
Futerra (2011) The greenwash guide. Futerra Sustainability Communications, London
Goodall C (2007) How to live a low-carbon life. Earthscan, London
Gore A (2007) An inconvenient truth: the planetary emergency of global warming and what we can do about it. Bloomsbury, London
GRI (2015) The global reporting initiative. https://www.globalreporting.org/. Accessed 7 June 2015
Hallman JC (2010) In utopia. St. Martin's Press, New York
Harvey M (2019) Utopia in the Anthropocene: a change plan for a sustainable and equitable world. Routledge, Abingdon
Henriques A, Richardson J (eds) (2004) The triple bottom line: assessing the sustainability of business and CSR. Earthscan, London
Hildebrandt S (2018) Verdensmålene - det vigtigste punkt på dagsordenen. KLS Pureprint, Copenhagen
Holdren J P (2018) A brief history of "IPAT" (impact=population x affluence x technology). J Popul Sustain 2(2)
ICIS (2019) About the International Centre for Creativity, Innovation and Sustainability. Available at: https://sites.google.com/site/iciscenter/about-icis/history. Accessed: January 2019
IPCC 6th Assessment Report (2022). https://www.ipcc.ch/assessment-report/ar6/. Accessed 1 March 2022
Kibert J et al (2012) Working Toward Sustainability: Ethical decision making in a technological world. John Wiley & Sons, Inc., Hoboken
Kirby J, O'Keefe P, Timberlake L (eds) (1995) The earthscan reader in sustainable development. Earthscan Publications Ltd., London
Klein N (2014) This changes everything. Simon & Schuster, US
Lynas M (2008) Six degrees: our future on a hotter planet, 2nd edn. Harper Perennial, London
McDonough W, Braungart M (2002) Cradle to cradle: remaking the way we make things. North Point Press, New York
Paris Climate Agreement (2022) UNFCCC. https://unfccc.int/process-and-meetings/the-paris-agreement/the-paris-agreement
Porritt J (2005) Capitalism as if the world matters. Earthscan, London
Princen T, Maniates M, Conca K (eds) (2002) Confronting consumption. MIT Press, Cambridge
Raworth K (2017) Doughnut economics: seven ways to think like a 21st-century economist. Random House Business Books, London

Rich N (2019) Losing earth: the decade we could have stopped climate change. Picador, London
Rio Declaration (1992) UN. https://www.un.org/en/conferences/environment/rio1992. Accessed May 2019, 5 Dec 2021
Roddick A (2001) Business as usual: the Triumph of Anita Roddick. Thorsons, UK
Rosling H, Rosling O, Ronnlund AR (2018) Factfulness. Sceptre, London
Sachs W (ed) (1992) The development dictionary: guide to knowledge as power, 5th edn. Zed Books Ltd., London
Sixth Assessment Report (2022) IPCC. https://ipcc.ch/assessment.report/ar/6
Sustainable Development Goals (2015) UN. http://www.un.org/sustainabledevelopment/sustainable-development-goals/. Accessed Dec 2015, 10 Jan 2022
The Brundtland Report (1987) WCED. https://sustainabledevelopment.un.org/content/documents/5987our-common-future.pdf. Accessed 9 Sept 2015, 14 Dec 2021
Wallace-Wells D (2019) The Uninhabitable Earth: A story of the future. Allen Lane, UK

Chapter 8
The Turning Has Started

Abstract This chapter highlights the challenging process of creating a state of global sustainability. It outlines the potential frameworks and factors needed and draws a trajectory for the future based on the sum of the diverse parts. In doing so, the chapter summarizes the key ingredients in the challenge to achieving sustainability and concludes with final reflections on its realisability. Questions relating to barriers and processes are addressed from the perspectives of the current state of sustainable development. Utopian theories and historical examples, as well as the practical experiences from the three cases are addressed. New influences are added, including recent global events and calls for action, and the impacts of the COVID-19 pandemic, the IPCC 6th Assessment Report from 2021/2022 as well as the agreements made at COP 26. The aspect of *inner sustainability* comprising the emotional, psychological, and spiritual dimensions, is examined as it plays a key role in the change of direction, as well as in building resilience to meet the new challenges as they arise.

8.1 The Ultimate Challenge

> We have not until now been faced with having to make a conscious, global decision on a path to take that includes and impacts the entire planet and all living species.

Sustainability is the ideal future political, social, and economic model for society. Multidimensional, global, national, and local factors all play a part in applying and achieving it. As such, it is the ultimate challenge for humanity. Sustainability is a complex and holistic construct—more so than any other societal model of the past. The impetus to generate this transformational change is more pertinent today than before. We have not until now been faced with having to make a conscious decision on a path to take that includes and impacts the entire planet and all living species. Scientific research alone is not enough to address the situation. Technology alone is not enough. Political intervention alone is not enough. Civil society's engagement alone is not enough. Pioneering initiatives alone are not enough. Business or financial initiatives alone are not enough because this transformational shift will require all actors at all levels of our global societies to intervene and collaborate. That is the ultimate challenge. Our species is called homo sapiens. But are we as a species

K. Blincoe, *Achieving Sustainability: The Ultimate Human Challenge*,
https://doi.org/10.1007/978-3-031-10023-9_8

advanced enough to secure the livelihoods of our own and future generations, and can we develop the necessary resilience, regenerate nature and live sustainably?

To answer this question—the possibility of achieving sustainability, I want to sum up the essence of the previous chapters, as they add to the mix of interrelated factors that are necessary in arriving at a relevant future pathway. There are no researched and tested 5, 10, or 12-step plans that we can follow to solve the problems. We can, however, learn and be inspired from history, utopian thinking, past crises, and from past and present initiatives such as the case stories I have documented in Chap. 6 and learn about the barriers that we need to overcome. All are important as they serve as parts in the kaleidoscope when drafting a framework and a master plan for the future.

We need that master plan.

Added to what we already know are other influencing elements, including the pandemic, its impacts and consequences, and the war between Russia and Ukraine, both are heralding a paradigm shift. They are showing us how we behave during an intense, global stress situation. Equally important are the recently updated scientific research findings on the impacts of climate change that are contained in the IPCC 6th Assessment Reports. They set the platform from which we can draw updated scientific data on what we can expect to have to deal with and what we must do. The summary of the agreements made at COP 26 in Glasgow in 2021 indicates how our leaders intend to address the situation and whether they have the will to do what is necessary. The first is positive; the latter is questionable. There must be a strong will and urge to act, no matter what. Where does this will come from, and how do we access the power behind it? The discussion touches on the aspect of what many call *inner sustainability*. I have in the book mainly focused on the outer or external aspects of the current predicament and what society can do to achieve sustainability. However, the personal *inner* dimension has a critical and pivotal role to play, as each of us needs to work with and support the transition regardless of how the process will affect us. The crucial test will be how we individually and together deal with it.

8.2 The Covid Effect

Since 2020 there have been countless reports on the impacts of the pandemic from north to south, east to west. From presidents to teenagers, doctors to patients, researchers to philosophers, professors to farmers, and families to communities. There have been countless interpretations, opinions, views, and thoughts on the nature of the pandemic and the way we have dealt with it. The unifying fact is that no one has had the answer, and no one has known how Covid-19 would develop and impact the world. We still do not know exactly where the virus originated. Was it made by humans, or did it emerge from wild animals? We do not know how it will continue to affect us or the exact consequences of it. History books will, in years to come, be able to document the global, national, and local impacts. But that is not possible today. We can only make intelligent guesses based on current research and experience gained. What we do know is the fact that the pandemic has directly or indirectly affected every human being on this planet. The virus has caused millions of deaths and endless suffering—and still does. In one year (2020) 2.2 million people died and millions suffered a severe illness, especially the elderly, and those who were vulnerable with underlying ailments and weaknesses. Indirectly, many more were affected, as people with other diseases were either not attended to or chose not to seek medical help.

When the pandemic hit the world, as we knew it stopped from one day to the next. Covid-19 disrupted the way of life everywhere, in the cities, in the villages and in rural areas. Public, as well as private life changed. Families were separated, travel was interrupted, businesses closed, and schools closed. The hospitals were struggling. The cities were devoid of people. Everyone had to stay at home. No celebrations, no sports events, no concerts, no cinemas, restaurants closed, shops closed, and religious worship in public places ceased. People died without their families being near and without proper funerals. Jobs were lost, and work was lost. The world went into *lockdown*—a new phrase that was spoken daily and everywhere. It was a crisis that came with no pre-warning and one which no nation had any experience in handling. It was unique, and even though a pandemic had been and still is predicted to be one of the ten greatest threats to humanity, we were unprepared for the devastation it caused.

Some of the long-lasting results of this crisis, both positive and negative, started to emerge during 2021. The young and the older generation were affected the most-the young from the absence of schools, friends, sports, and social events. Many subsequently suffered from mental disorders, mainly in the form of depression. The elderly had a particularly raw deal, as they were vulnerable and fearful and unable to see family members or friends. Women and girls were exposed to increased domestic violence and sexual abuse by male family members. UN Women (an entity in the United Nations) stated that since the outbreak of the virus, data showed that violence against women and girls, particularly domestic violence, intensified and increased (UN 2022). Cultural institutions, tourism, and people working in these fields world wide suffered financial and job losses. The virus caused a melt-down of large parts of the world economy, as the illness sparked an immediate recession followed by a supply crisis. Luckily, many damaging consequences were alleviated by governments intervening on a global scale.

For some, there has been positive changes. One major benefit has been the opportunity to work from home as most of the workforce was ordered to stay away from the office or workplace. People had to be able to plan a schedule for the day involving normal working hours, household chores, and playing, teaching, and supervising the children 24/7. As the impact of the virus eased, many people maintained their home offices but went to work 2–4 days a week. A love for walking, enjoying nature, home cooking, gardening, knitting, and many almost forgotten skills and activities were revived and are still maintained.

8.3 End of Neoliberalism?

> New development theories based on the old do not suffice in this context. A more drastic transition agenda is needed - our imagination and inventiveness are needed.

The political and economic paradigm changed during the pandemic. There was a significant shift from market-driven decisions of the neoliberal era to citizen-orientated political choices of a more Keynesian nature. Governments stepped in and took the measures they deemed were right, both regarding the behaviour of citizens in the form of lockdown of public and private places, stipulating masks, tests, and subsequent vaccination programmes, as well as offering financial support for businesses. Governments chose to introduce new policies dealing with resource distribution, tax cuts, and public spending, supporting the failing industries and their employees to a level not seen since World War 2. Suddenly lives mattered more than profit, collective action mattered more than individual action, and government intervention mattered more than the reliance on supply and demand. Globalisation came into question. Countries closed their borders and started to look inwards. The narcissistic *me* was substituted by a necessary new *us*. Another impact was the change in consumerist patterns. Shopping went online storage and transportation industries grew exponentially and made Jeff Bezos of Amazon the second richest person in the world, followed by the owners of other main online transportation, and shipping industries. Entertainment such as Facebook, Instagram, Twitter, Pinterest, TikTok etc., expanded exponentially, unfortunately promoting quantity rather than quality.

The lingering effect of the political measures and people's behaviour points to a shift in both politics, economics, and social infrastructure. The question is, as the economy is picking up, whether the pandemic has ended the era of neoliberalism, as many maintain, and given birth to a new more socially and environmentally orientated model, or whether there will be an attempt to return to the pre-pandemic norm. During 2020, people started to talk about the opportunity to rebuild societies in a new way. The 'EU Next Generation' and the 'US Build Back Better' programmes were some of the new agendas launched by the governments. On the sustainability side, the US introduced the 'Green New Deal', Europe launched the 'European Green Deal', and the UK put forward proposals for their 'Green Deal'. Many started to call for a new *social contract*.

During the pandemic, there was an impetus to create an environment of positive and green initiatives. Suddenly inhabitants of the smog-filled cities discovered the blue sky above the buildings. Others could see landscapes and mountains that had been hidden from view. Nature became more present and was more appreciated. People felt claustrophobic in their skyscraper flats as opposed to those living in the country with access to open fields and walks in nature and along waterways. So, the desire for city living did seem to diminish, at least for a while. Even though some economies appeared to be recovering, nothing was the same. Several new threats started to loom, such as rising energy prices, increasing inflation, and supply-chain bottlenecks. A wide range of production and building materials, food and household products, and fashion became more expensive and hard to come by.

Many believe that politics and economics have been reshaped, and globalisation has come into question. Countries are looking at new economic models with which to face an uncertain future. Democracy is being challenged by autocratic and nationalistic tendencies, and tensions are building between the major powers of the US, China, and Russia. There is a growing feeling among people of anxiety and fear. However, there are also contradictory developments. Civil society's values are being challenged by extreme populist expressions at the same time as self-help activities and initiatives are flourishing. A new type of activism is rising aimed at politicians, protesting against their inaction to deal with the crises and the causes, and youth movements have been launched such as Fridays for Future, and Extinction Rebellion.

When the pandemic seemed to be relatively under control, a war broke out between Russia and Ukraine, initiated by Russia. From a European and an American point of view, it caused an uproar, and a massive strategy of sanctions was planned and imposed on Russia as a response. The atrocities incurred by one country on another are almost impossible to comprehend but it tells us that humanity is still at a primitive stage of evolution. Apart from the direct and devastating suffering inflicted on the people of Ukraine, the war will, of course, impact the rest of the world both directly and indirectly as a result of the sanctions and destruction of a resource-rich country in terms of both wheat, sunflower oil, and other commodities such as metals and minerals. Already, there is a deepening resource and food crisis covering a variety of necessities. Inflation continues to increase, and the financial markets are jittery. Adding the underlying effects of the pandemic to those of the war, the sum of the two could point to a collapse of the current economic framework. From a sustainability viewpoint, this could also pave the way for a new different, fairer, and more holistic economic model. The EU has been dependent on oil and gas from Russia. This will now have to change. The golden opportunity is to invest massively in green technologies, and politicians in Western countries have promised to escalate the production of renewable energy.

8.4 Net Zero Transition

The current situation is, of course, further stressed by the climate change impact, already causing havoc around the world. The recent Global Risks Report (2021) stresses the fact that embarking on reducing CO_2 emissions at the scale required globally might cause what the report calls a "disorderly" net zero transition, where the impacts of the needed interventions could cause "economic volatility", and destabilize the financial system. There is pressure on governments to act fast, and this will very likely result in policies and regulations that either aren't sufficiently researched and tested, are getting implemented too fast, or too late, or will not achieve the right results. This will cause a variety of problems from job losses, lack of the right skills, lack of resources, supply problems, and shortages. The Global Risk Report tells businesses to expect a chaotic transition process, and suggests that they embrace it, drive innovation and new ideas in order to survive as well as benefit from the process. The report compares this scenario to the pandemic where the businesses that were agile, flexible, and ready to provide what was needed survived and emerged unscathed from the crisis. A new type of entrepreneurship emerged, establishing small and innovative businesses that are flexible and agile, and can meet new and different demands.

It is logical and realistic to assume that the change to a net zero future will cause disruption. It cannot be avoided. Achieving zero CO_2 emissions will change everything. Achieving a state of sustainability will change everything. By the same token, there will also be disruption and chaos if governments decide to continue to promote *business as usual.* Either way, we are in for a rough ride. If governments had embarked on the transition when the first researched reports on climate change, biodiversity loss, pollution, and other negative effects started to emerge and affect our lives several decades ago, there would have been time to course correct. It could have been done through proper planning, consequence assessment, research, experiments, prototyping, and tests. However, as no one knows for sure what a zero-carbon transition will entail, and as there is little time for experiments, the prospects for the future are unpredictable.

The latest IPCC Summary for Policymakers (Working Group II report, Climate Change 2022: Impacts, Adaptation and Vulnerability) has estimated that working towards low emissions and climate resilient development still has a chance if applied now. The gap has narrowed to three years to make the required impact. The report also states that the dangers we are facing now due to the weather extremes will be accumulating and cascading, meaning that sudden changes in one area might lead to sudden changes in other areas causing a domino effect. Some of these changes can no longer be alleviated. We must, therefore, do what we can to adapt to climate change, cut emissions, build resilience and survival mechanisms, and simultaneously embark on the road to sustainability.

The Global Challenges Report from 2021 names weapons of mass destruction as the first major threat to humanity, and climate change and ecological collapse as the second and third most serious threats. The latter two include the changing weather patterns, the collapse of ecosystems, the decline in natural resources (water in particular), and the chemical pollution of the Earth's systems. These are human-induced, and the failure nationally and internationally to act preventatively on these risks must be the greatest threat of all. The IPCC Reports tell us that environmental, social, and economic factors should be addressed simultaneously and that the most important act is to safeguard and strengthen nature. To do this, all stakeholders in all sectors of society, such as the public and private sectors, academia, and the financial sectors, must be involved. We need policymaking, research, technology, and financing to be integrated into an overall symbiotic plan for relevant and specific interventions and make sure that there are flexible course corrective actions and contingencies on stand-by.

The positive part is that the world is stepping up on initiatives. The UN Newsletters are reporting on numerous positive undertakings, such as the Race to Zero campaign, where they have developed frameworks for reducing emissions by 50% within the next ten years in over 30 sectors. There is a flurry of organisations and expert groups, including cities and regions, businesses and financial institutions around the world, that are increasing their endeavours to collaborate across countries, sectors, and multidisciplinary fields.

8.5 Promises, Pledges and Agreements

The aims of the UN Convention COP 26 in November 2021 were, for the world leaders, to address the escalating problem of global warming and the changing climate and to be seen to act preventatively. The results were therefore interesting to record. What action were the countries and global stakeholders proposing to take to mitigate and adapt to climate change, and other environmental problems. The additional perspectives that made this COP even more significant than the previous 25 Conventions were the impacts of the Covid crisis and the latest research findings from the 6th Assessment Report from the IPCC. The initiatives from previous COPs had not had the desired effect, and some nations, especially the richest and largest countries reneged on proper and binding commitments. This was evident at COP 26 were neither the Chinese nor the Russian leaders showed up, and where China postponed the promise to cut emissions until 2060 and India until 2070. These governments, backed by a few others, were able to change the final version of the Glasgow Climate Pact in their favour in terms of coal production and use (I also mention this in Sect. 2.4 in Chap. 2). These nations clearly showed us that they had other priorities and that climate change was a secondary issue for them as much can happen politically, economically, and on the weather front before 2060 and 2070. A common denominator between all nations at COP 26 was that embarking on fulfilling or meeting the commitments, promises, and pledges was placed somewhere in the future. That was for me the most troubling aspect of the results of the convention. No one said we will do this, starting today. Acting on the increasing urgency was as elusive as in all previous COPs, although there was more awareness at this event.

The Race to Zero programme is challenging, but as the Covid-19 pandemic and economic crises have shown, these obstacles can be overcome and dealt with. The hope is that the countries resisting will be pressurized to shift their focus through trade deals, contracts, finance and technology, as well as psychological pressure from civil society. On the positive side, new agreements were made, and many partnerships and alliances were set up. Goals were clarified. Intentions flourished. It seemed that globally we are more informed of climate change and its effects than ever before. We are all impacted, and more and more actors from all walks of life are joining the fight to mitigate and adapt to the changes. These include organisations representing indigenous people, nature, cities, culture, religions, and youth-orientated groups.

Some of the specific outcomes from COP 26 consisted first of all, and most importantly, of the Glasgow Climate Pact. Secondly, the pact included agreements such as a requirement for world governments to report annually on updated NDCs

(Nationally Determined Contributions). This was the result of The Action Tracker (CAT), a climate analysis coalition, providing an updated assessment and predicting a rise of 2.4 °C based on the countries' short-term goals for the next decade. It was, therefore, important to revise the NDCs annually, rather than every five years, which was the original decision. The first time this will happen will be in Egypt at COP27 in 2022. Thirdly, to meet the target of 1.5 °C it was decided that coal, the dirtiest of all the fossil fuels, had to be phased out starting immediately. However, as many countries including China and India still rely on oil and coal the final agreement in the Glasgow Climate Pact was changed to a *phasing down* of coal production and usage rather than a *phasing out* commitment. Other agreements dealt with economic issues. One agreement was a change from the commitment in 2009, to help poorer countries cut their CO_2 emissions. This included an increase in the contribution to $500 billion over the next five years toward mitigation initiatives, as well as a doubling of the financing of adaptation initiatives. Many financial institutions agreed to address agriculture-orientated deforestation, and a partnership between 500 million farmers and the Regen 10 initiative promised to ensure that over half of the world's food production will be sustainable by 2030. In addition to the above, other pledges were made. One to stop deforestation. Another to provide public and private finance to pay for the transition to renewable energy as well as the reduction of methane gas. Lastly, to build a global fleet of electric cars and advance renewable energy technology worldwide. Many deals were made, and partnerships formed, which if acted upon, would further a worldwide CO_2 reduction.

The feedback from COP 26 was mixed. The rich countries believed they had done what they could—or wanted to. The poor and developing countries felt that the agreements in Glasgow were not extensive enough and exposed them to a very vulnerable future. Disappointment and anger were expressed as the Climate Pact failed to include promises to finance the *loss* and *damage* to countries caused by climate disasters. The stance was that as global warming was mainly caused by the rich countries, they must pay for the rebuilding and reconstruction of the affected countries and compensate them for the loss of livelihoods and jobs. The lack of inclusivity of members from the civil society sector was also a key point in the criticism that followed the event. Many were barred from attending due to restrictions. It was felt that only the elite was allowed to attend vital discussions.

COP 26 was a platform that dealt with challenging issues, including the financing of global interventions. An entire palette of actions was offered that, if enforced, would help meet the ambitious goals mid-century. If the promises are kept, the pledges are honoured, and the commitments are met, the aim to reach 50–70% reductions in emissions by 2030 and zero emissions by 2050 might well be possible. However, there is no guarantee that the world governments and other actors, especially those dealing with business and finance, will keep their word. Despite this, I believe a *turning* has begun, however slowly—but nevertheless, it has begun as a consequence of the sum of factors I have outlined in the previous sections. Once the wheels are turning, they will hopefully gain speed.

8.6 The Will to Change

The Danish newspaper "Information", reported in their leader column (16th February 2022) on the recent recommendations from the Danish Climate Citizen Council (Klimaborgerting), which was formed in 2019 as part of the Climate Law. The establishment of the Council was based on international findings that have seen such councils making positive and constructive contributions. The Council consists of 99 people, made up of a random selection of citizens from all over the country. The role of the Climate Citizen Council is to recommend solutions to the climate change challenges. The first set of 114 recommendations was delivered in April 2021. The second set was finalised in February 2022 and consisted of 73 recommendations. A significant outcome of the second set of recommendations was that the members of the Council believed that a reduction in resource use was more important than continued economic growth. Similarly, they agreed that a Danish contribution to CO_2 reduction to keep global warming at 1.5 °C or below 2 °C was also more important than economic growth. A majority of members wanted the political parties to jointly take responsibility and tackle the climate crisis with as much resolve as they showed in the pandemic. The members demanded action here and now, stating that economic interests must not take precedence. The article about the work of the Climate Citizen Council included interviews with the various members. What I find significant in what they said was their surprise and shock at discovering that many political interventions, that could have helped alleviate the current problems decades ago were not made. It illustrates that the general lack of information and knowledge of ordinary people regarding these problems has been a contributing factor to the current crises. It also demonstrates the disconnect between politicians and citizens, and the danger of not-knowing.

The sentiments and recommendations relayed above were from ordinary people who did not work with climate or sustainability issues on a day-to-day basis. As climate crisis intensified and greater awareness spread, people in the mainstream reacted in different ways. Most carried on with their lives as usual. For some, the increased awareness made them talk about the issues at home and on social media, and many changed to living and consuming differently, such as buying more organic food products, more second-hand clothes, and used furniture. They became more aware of waste and increased interest in recycling. Some people, in particular the younger generation, have become very passionate about climate change and their challenging futures. Many next generation networks and groups have been set up that mourn the past and the loss of species, and they worry about the foundation on which their lives depend. As in any other healing process I believe that facing and accepting the reality of the past is crucial, as it encourages new thinking. I equally believe that facing the future with creativity and imagination is necessary to invoke the needed change. These points are discussed further in the text below.

Inner Sustainability—Inner Resilience

The cause of climate change has mainly been connected to economic and political activities. But many also place the responsibility on the individual, which can be a heavy burden. Being and staying positive with the knowledge that most of us now has regarding the reality of the current and future situation can be a challenge. Many young people suffer from climate anxiety and fear of the future, and elderly people feel remorse. I do not believe that the ordinary person is to blame for the state of play. I have in the book mainly blamed the state, the governments, and the politicians. The reason is, that if you have the knowledge and deliberately choose to ignore the facts, that are a detriment to the people you are supposed to protect, you are both responsible and accountable. Future generations will not look back with kindness on the politicians or governments of the Anthropocene.

The inner environment is as important as the outer. It has just not been given as much attention, unfortunately. One reflects and mirrors the other. Inner turmoil has a negative impact on your life, both personally and professionally. It affects your state of health and well-being. Given that we have embarked on an uncertain, insecure, and volatile journey regardless of which path we take, it would be wise to build up our inner resilience to withstand the outer turmoil we have to go through. Humans function on different levels, such as the mental, the emotional, the physical, and the spiritual (there are more levels, but these are the basic ones). A sage wisely said that to live a balanced life, we should visit each of the *four rooms* in this *house* daily. From my perspective, the two most important elements (or *rooms*) to build inner resilience are working with the emotional and the spiritual aspects of our lives. Emotions, negative as well as positive, tend to control our actions, and a lack of attention to our spiritual needs can create a sense of emptiness, meaninglessness, and depression.

There is a new initiative called the Inner Development Goals project in Sweden. The Inner Development Goals was created by a group of academics and researchers in 2019 to address the lack of action on the 17SDGs and on the sustainability agenda in general. They decided to explore the kind of qualities, abilities, and skills that should be present both in individuals as well as in organisations and groups to fulfil the visions and meet the goals. Their focus is on the inner qualities and competences necessary to shift the impasse. The group has conducted an extensive survey on the issues. It has, so far, narrowed the focus to five states of being: Being (relationship to self), thinking (cognitive skills), relating (caring for others and the world), collaborating (social skills) and acting (driving change) (IDGs 2022). It is clear that the psychological dimensions matter much more than we think. The gap between talk and action is wide when dealing with the implementation of sustainability. There is resistance to bridging the gap and moving the agenda forward, and it is the resistance and procrastination that need addressing. Nearly everyone I speak to is worried, depressed, or feels a sense of hopelessness. There is a lack of positive visions and scenarios. Positive dreams and imagination play a role both in achieving sustainability but also in creating resilience. That is an area we can address.

Other organisations and networks are working with inner aspects, for example, the Deep Adaptation network which is a UK-based social movement and network set up by the English Professor, Jem Bendell, previously mentioned in Sect. 7.3. He worked with and researched sustainability for many years and developed the view that the potential for a societal collapse was present and that humanity needed to prepare for this. I have experienced one of Jem Bendell's seminars, where the audience was encouraged to feel the emotions in a context of a collapse. It was a very emotional experience. The belief is that addressing the emotions will lead to becoming more resilient and being able to adopt values such as non-violence, compassion, respect, and empathy for positive action. Steffi Bednarek, a climate psychologist, said in an interview (in 2022) that "what we don't feel is unlikely to lead to any fundamental change in our behaviour", meaning that we tend to shut off our feelings when dealing with climate change because we cannot deal with it. We are part of a global system that is causing damage, and we shut off because we cannot bear to face the consequences of what we have done and are doing. Bednarek talks about compartmentalisation and that we in the Western world are brought up and taught to compartmentalise our emotions in order to survive. This view resonates with the fact that we are becoming more and more separate and cut off from meaningful and responsible interactions with our natural habitat, with what sustains us—and in many instances, with our fellow human beings. This generates a deep sense of loss and a sense of loneliness, unhappiness, restlessness, or depression, and we numb it with drugs, alcohol, stressful work, abuse, and obsessive and violent behaviour. Depression is currently on the rise, and some believe that this is a globally felt depression, which with all its negativity, is spreading in the same way as the pandemic. As we are all connected and belong to one species, I believe that this could be so. When depressed, it is difficult to be positive, powerful, full of joy, energetic, and ready and willing for change. This state can then lead to further negativity and powerlessness in a downward spiral that becomes difficult to shift.

Getting to that point, that state of powerlessness has happened to me from time to time in my 30+ years in the sustainability field. But I have learnt to return to that meditative place in nature that I found in my childhood (described in the Preface) or to a similar joyful place. A place where I can immerse myself in nature, contemplate a ladybird, a tree, a snowdrop, a butterfly bush, insects in the lavender, or the wild geese flying overhead. Joy and happiness return, with that, the energy to take the next step and do what I do. If you do not have nature outside your front door, then listen, for example, to the voice of the Buddhist monk, Thich Nhat Hanh (1926–2022), who may simply ask you to observe your breath. That does it too. Meditation is an excellent tool for this. Brahma Kumaris mentions meditation as a lever to achieve sustainability. Others have other ways, and you will find your way. We are lucky, that today there is a remedy somewhere for most inner and outer disorders or imbalances. For inner well-being, there are numerous practitioners accessible online or offline to talk to and learn from. I said at the beginning of this section that it is important to face and take responsibility for the past. However, remaining and dwelling in a state of guilt, shame, and feeling bad will not help us move on.

8.7 The Challenge

Sustainability is about ethical and responsible behaviour. It is a state of the world where people can lead decent and dignified lives within the carrying capacity of the earth no matter who they are or where they live. That means exercising responsible stewardship. It sounds logical. No conscious, intelligent, or wise person would destroy the environment that is feeding and providing us with all and more than we need. This is common sense. Everything the human species has created over the past two hundred years until now has been based on the assumption that planetary resources are unlimited. There has been little or no regard for the possible negative consequences that such an attitude could bring about. Despite available facts on the long-term effects of global warming, which were already known in the late 70s, *willed blindness* set in, which was spread by governments and the oil and gas industry in the name of profit-taking and wealth creation. I can recommend that you read the author and New York Times Magazine writer Nathaniel Rich's book "Losing Earth: The Decade We Could Have Stopped Climate Change" from 2019. He narrates that in the early 1980s, American and other international scientists predicted that there would be significant global warming over the coming decades generated by the fossil fuel industry and that if this was not dealt with, "it would be too late to avoid disaster". The world was alerted, and a treaty to reduce carbon emissions, supported by 60 nations, was ready to be signed. But America, already the largest CO_2 emitter, reneged on the agreement, so industrial development continued at an increasing speed—in the name of the *public good*. This brings us to today, where we are now faced with what was predicted then. The impacts, we now know, will increase in size and intensity. The *willed blindness* made governments, industry, commerce, and finance resist mitigating when there was time. When the UN introduced the sustainable development agenda in 1987, many scientists already knew what we were up against. I met some of them when I was the Director of Schumacher College (2006–2008), a college for sustainable learning in Devon, UK. When I asked what they thought of the future, most declined to answer. Not because they did not know, but because the answer was too depressing, and I have over the years met several professors in environmental departments in universities who have either become cynical or deeply concerned and depressed with the status quo.

Governments have launched many important initiatives to implement sustainable development, such as setting up ministries for the environment and climate and have implemented various policies aimed at addressing the crises. The global conferences COP 20 in 2015, where the Paris agreement was signed, and COP 26 in 2021 have forced governments to commit more seriously to curb CO_2 emissions. We need to push this interest forward and make sure their promises are kept. We need to revisit utopian thinking and turn to history and learn about what was done and what can be done to propel the turning upward and keep it moving towards a state of sustainability.

An Abundance and Sufficiency Model

> Utopias are always radical rather than reformist. They argue that what's needed is not just a modification of the existing system but a completely different system (Harvey 2019, p. 6).

What we are looking for is not a Thomas More 'Utopia', nor is it a 'paradise' based on religious beliefs. It is a practical, down-to-earth utopia set within a specific time and place, here and now or in the near future. It is not in a far-away place separated from the mainstream. I want an all-encompassing model where people can have contented and meaningful lives without the continuous threat of hunger, poverty, disease, violence, and abuse. A model that balances human and nature's well-being. It will be a model where the 17 Sustainable Development Goals are met while respecting individual cultures and traditions. A model that can and must be adapted to fit every continent, every nation, and every society. A social, environmental, and economic master plan for sustainability, that is effectuated by the state, backed by societal actors, and supported by public and private finance.

As an example of large-scale activities, Denmark, last year, launched a series of CO_2-reducing initiatives, including building wind-energy islands intended to supply fossil-free energy to ten million homes. The islands are intended to produce three gigawatts of electricity to be increased to ten gigawatts over time, which would amply cover the country's energy needs and make export possible. As a response to the new war-related energy crisis, the Danish government launched a new agenda called "Danmark kan mere II" (Denmark can do more II), that eventually, will make the country independent of Russian gas and will move us towards reaching the 2030 and 2050 goals for reduction of CO_2 emissions. This time the government is not waiting for new, not yet invented technologies but is promoting the use of already known methods and models to achieve the goals. The agenda includes the substitution of all gas and oil-fired central heating systems with heat pumps to be finalised by 2028. New energy islands will be built in addition to the two already launched, and four times as many windmills and solar panels will be established on land. The government has also introduced a taxation on CO_2 emissions in the industrial sector. There is a push for electric cars and the development and production of Power-to-X conversions of electricity, storage, and reconversion pathways. It is the intention to use the new energy islands as a base to produce green hydrogen from seawater and to start protect the sea and waterways from further pollution, and restore their health. These are ambitious initiatives and tell us that the impossible might be possible given the right impetus.

To create the modern-day utopia, we must invent one that is both a utopia of sufficiency and abundance (Sect. 4.4). The sufficiency utopia embraces non-material, human, and nature-related aspects of life, some of which can be measured and quantified. The sufficiency utopia is the model where life is based on simplicity. It is community orientated, with a focus on nature. Production provides the basics of life, and waste is not an option, similar to the values and principles stated in the Brahma Kumaris case story (Sect. 6.5). The Brahma Kumaris Environment Initiative is based on five main principles: Living with simplicity, buying compassionately, using economically, learning continuously, and sharing generously. There are other values in the sufficiency context, such as beauty and aesthetics, that enhance people's lives and are important for human well-being. The field of abundance consists of the application of science and technology to support sustainability endeavours. Models, such as Circular Economy and Cradle-to-Cradle, are part of this segment, as are most of the models, methods and assessment tools described in Chap. 5. New technological advances are essential to create and realise the modern-day utopia. To achieve the needed goals we need both fields supporting the endeavours. Even if the 17 SDGs represent the sustainable development framework closest to a unified model of sufficiency and abundance, it lacks cultural content such as the arts. It also lacks elements such as aesthetics, beauty, and other *inner* concepts, including those of a spiritual origin. As previously mentioned (Chap. 4), many small intentional communities exist that have experimented with creating the utopia of their dreams, and many have succeeded. In a conscious drive to recreate or redesign our societies to become sustainable, some of these intentional communities can act as models, prototypes, or test cases for larger-scale implementation.

Barriers Can Be Overcome

There are many barriers to sustainability. I have devoted a chapter (Chap. 7) to describing some of those I discovered in my research and those uncovered in the case stories. If the political aim and focus are on implementing sustainability most of the barriers can be overcome through a top-down driven intervention such as the passing of new laws, policies, and legislation. One of the key lessons from the island of Samsø (Sect. 6.2) was the importance of inclusivity, making people part of generating solutions. Upscaling this element to a larger context, a method to include citizens could involve conducting a referendum to ask the electorate to give the government the necessary powers to appropriately address the crises and implement a master plan. In addition, to include citizens in developing and deciding on the solutions and the actions and interventions, that must be made to implement the plan.

An inspiration to this is the way Nelson Mandela, who was the first black President of South Africa from 1994–1999, introduced the *truth and reconciliation* measures to deal with the painful past of apartheid. It was an almost impossible task, considering the dramatic and traumatic events the South African people had lived through. Professor Don Beck, an American geopolitical adviser, and theorist was brought in as a consultant. He developed the theory behind a value-based model for the evolution of businesses, organisations, and societies, with his colleague Christopher Cowan called Spiral Dynamics. This work is based on the work on cyclical theory on adult human development by professor of psychology Clare W. Graves (1914–1986). I heard a lecture by Don Beck in Denmark in the 90s where he talked about Spiral Dynamics and explained the principles of evolution in the context of this model. The system is based on human values, and deals with how ideas and behaviour spread through culture—and how you can initiate changes to and in a culture through the application of the model. The principles are described in the book "Spiral Dynamics in Action: Humanity's Master Code" written by Don Beck and colleagues and published in

2018. Don Beck was an important advisor to the government in the South African transition in 1994 from a white to a black rule. He used Spiral Dynamics to bring together opposing factions and set up co-creation and transdisciplinary meetings to ensure a non-violent transition before and after the elections in close collaboration with Nelson Mandela and Archbishop Desmond Tutu. Don Beck planned the World Cup team of South Africa in 1995 with Nelson Mandela, which was a crucial element in uniting the country. It was an extraordinary achievement that seemed an impossible task to us on the outside. Unfortunately, we have not been as wise as Nelson Mandela, but it is examples such as these that can inspire us to achieve what seems to be the impossible. South Africa is also one of the only countries that, when it changed its constitution in 1996, introduced protection of the environment. Section 24 in The South African Constitution and Bill of Rights states:

> **Section 24: Environment**
>
> Everyone has the right to:
>
> An environment that is not harmful to their health or well-being
>
> Have the environment protected for present and future generations.
>
> The government must pass laws that prevent pollution and damage to our natural resources, promote conservation and make sure that natural resources are developed while also promoting the economic and social development of people.

Each of us has rights. These rights are given through the constitution and laws of the country where we live, and they differ from country to country. The rights of people in democracies are different from those in autocracies. I have, for example, the right to vote, free education, medical treatment, a pension when I am 67, and much more. In autocracies, people have fewer rights. If the young generation wants the right to clean water, air, and soil, as many now demand, they must manifest that right through electing the right people to government, who will pass the laws that give these rights—or through activism demand that the government changes the constitution.

All the ingredients are available that are needed to address the crises, overcome the barriers, and move society towards sustainability. That includes technological know-how, financing, models, methods, and processes. The case stories and other examples of implementation initiatives suggest that the possibility of achieving sustainability exists. History has shown that crises can be overcome. I believe that the barriers to change, that matter are primarily cultural and ethical. We have the knowledge, the information and technology needed—and the finances. The positive aspects are that many actors are coming together worldwide in partnerships, alliances, and collaborations, such as those that resulted from COP 26. However, the crucial fact is that despite these positive aspects, CO_2 emissions are still rising.

The latest IPCC Report (2022) is bleak, to say the least, and one thing is certain: we are faced with an extremely uncertain and volatile future. Not only regarding climate change but also regarding biodiversity, chemical toxicity, pollution, limited resources, and other factors that are a risk to our human existence. This is the challenge. If the promises made at COP26 are kept and met, we are in for a 10–30-year period of deep transformation of our societies and our lives, where the fundamental systems and infrastructures underpinning the current way of life will be replaced. This will be uncomfortable. If, on the other hand, the world governments renege on their promises and try to continue to keep to the status quo, we are also in for a period of massive and possibly unsurmountable challenges as the impacts of the universal imbalances deepen and intensify. During the past two years, we have been exposed to a pandemic, droughts, locusts on the African continent, raging fires in Australia, Canada, and the US, extreme heat in Canada, Greenland, and Europe, floods in Europe and Asia as well as hurricanes across the globe. The turning has already begun, and the spiral can move downward or upward depending on what we do next. This is a crucial time.

To succeed, we need institutional imperatives to act on and support. We need a redesign of the political and economic landscape, where the state takes responsibility side by side with its citizens to right the current wrongs. Turn the tables and start to rebuild the broken parts—create a global Marshall plan and/or call a referendum with the aim not only to ensure collective welfare but to regenerate nature, restore biodiversity, minimise pollution and waste, and reduce our interference with the natural habitat. Achieving sustainability is the ultimate human challenge, and we already have everything we need to make it happen. Let's rebuild and protect the garden that was entrusted to us.

The Wind, one Brilliant Day

The wind, one brilliant day, called
to my soul with an odor of jasmine.
'In return for the odor of my jasmine,
I'd like all the odor of your roses.'
'I have no roses; all the flowers
in my garden are dead.'
'Well then, I'll take the withered petals
and the yellow leaves and the waters of the fountain.'
The wind left. And I wept. And I said to myself:
'What have you done with the garden that was entrusted to you?'
(Antonio Machado 1875–1939).

Readings and References

About Brahma Kumaris (2015–2022) Brahma Kumaris. http://www.un.brahmakumaris.org/. Accessed 2015–2022

About Energy Academy (2015–2022) Samsø Energy Academy. http://www.energiakademiet.dk/en/. Accessed 2015–2022

About Novozymes (2020) Novozymes. https://www.novozymes.com/en. Accessed: 2015–2022

Beck DE, Cowan CC (2005) Spiral dynamics: mastering values, leadership and change. Blackwell Publishing

Bell S, Morse S (2008) Sustainability indicators: measuring the immesurable, 2nd edn. Earthscan, London
Benyus JM (2002) Biomimicry: innovation inspired by nature. Harper Perennial, New York
Blewitt J (2008) Understanding sustainable development. Earthscan, London, p 157
Blincoe KS (2020) Essential sustainability: from utopian theory to practical application. Dissertation, Kingston University. https://eprints.kingston.ac.uk/id/eprint/51146. Accessed 2020–2022
Braden G (2014) The turning point: creating resilience in a time of extremes. Hay House UK Ltd., London
Brundtland Report, Our Common Future (1987) UN. https://sustainabledevelopment.un.org/content/documents/5987our-common-future.pdf. Accessed 10 Jan 2021
Build Back Better Act (2021) US Congress. https://www.congress.gov/bill/117th-congress/house-bill/5376/text. Accessed 3 Dec 2021
Caradonna JL (2016) Sustainability: a history. Oxford University Press, New York
Carey M (2021) Value(s): building a better world for all. William Collins, Croyden
Circular Economy (2015) Ellen Macarthur Foundation. http://www.ellenmacarthurfoundation.org/circular-economy. Accessed 5 April 2015
Climate Action Tracker (2022). https://climateactiontracker.org/. Accessed 17 March 2022
COPs (2022) Conference of the parties. https://unfccc.int/process/bodies/supreme-bodies/conference-of-the-parties-cop
CSR (2015) CSR. http://csr.dk/de-gode-penge-ligger-i-b%C3%A6redygtighed. Accessed 15 Sept 2015
Danish Climate Citizen Council (2022) Information
Diamond J (2011) Collapse: how societies choose to fail or succeed, 2nd edn. Viking Penguin, New York
Diamond J (2012) The world until yesterday. Allen Lane, London
Doppelt B (2003) Overcoming the seven sustainability blunders. Syst Thinker ® 2(14):5
Dyer G (2011) Climate wars: the fight for survival as the world overheats, 2nd edn. Oneworld Publications, Oxford
EU Green Deal (2022) European Commission. https://ec.europa.eu/info/strategy/priorities-2019-2024/european-green-deal_en
EU Next Generation (2022) European Union. https://europa.eu/next-generation-eu/index_en. Accessed 25 Jan 2022
Exstinction Rebellion (2022) Rebellion. https://rebellion.global/. Accessed 19 April 2022
Farley HM, Smith ZA (2014) Sustainability: if it's everything, is it nothing? Routledge, Abingdon
Futerra (2011) The greenwash guide. Futerra Sustainability Communications, London
Glasgow Climate Pact (2021) UNFCCC. https://unfccc.int/process-and-meetings/the-paris-agreement/the-glasgow-climate-pact-key-outcomes-from-cop26. Accessed 5 Jan 2022
Global Challenges Report (2022) Global challenges foundation. https://globalchallenges.org/initiatives/analysis-research/reports/. Accessed 12 April 2022
Global Risks Report (2021) World economic forum. https://www.weforum.org/reports/global-risks-report-2022. Accessed 8 Dec 2021
Goodall C (2007) How to live a low-carbon life. Earthscan, London
Gore A (2007) An inconvenient truth: the planetary emergency of global warming and what we can do about it. Bloomsbury, London
Graves CW, Cowan C, Todovic N (2005) The never ending quest: Dr. Clare Graves explores human nature: a treatise on an emergent cyclica. ECLET Publishing
Green New Deal (2021) US Congress. https://www.congress.gov/bill/117th-congress/house-resolution/332. Accessed 5 Dec 2021
Hallman JC (2010) In utopia. St. Martin's Press, New York
Hanh TN, McLeod M (ed) 2010 you are here: discovering the magic of the present moment. Shambhala, Boston & London
Harvey M (2019) Utopia in the Anthropocene: a change plan for a sustainable and equitable world. Routledge, Abingdon

Hildebrandt S (2018) Verdensmålene - det vigtigste punkt på dagsordenen. KLS Pureprint, Copenhagen
Holdren JP (2018) A brief history of "IPAT" (impact=population x affluence x technology). J Popul Sustain 2(2)
IDGs (2022) Inner development goals. https://www.innerdevelopmentgoals.org. Accessed 20 March 2022
IPCC 6th Assessment Report (2022). https://www.ipcc.ch/assessment-report/ar6/. Accessed 1 March 2022
IPCC SPM (2022) IPCC. https://www.ipcc.ch/report/ar6/wg3/. Accessed 29 Feb 2022
Kibert J et al (2012) Working toward sustainability: ethical decision making in a technological world. Wiley, Hoboken
Kirby J, O'Keefe P, Timberlake L (eds) (1995) The earthscan reader in sustainable development. Earthscan Publications Ltd., London
Klein N (2014) This changes everything. Simon & Schuster, US
Klimaborgerting (2022) Information, leder: Haves: 73 anbefalinger. Handlekraft NU, Ønskes. https://www.information.dk/indland/leder/2022/02/haves-73-anbefalinger-oenskes-handlekraft. Accessed 17 Feb 2022
Limits to Growth (2019) Club of Rome. https://www.clubofrome.org/news/publication-of-the-limits-to-growth/. Accessed 10 Oct 2019
Lynas M (2008) Six degrees: our future on a hotter planet, 2nd edn. Harper Perennial, London
More T (2010) Utopia. Feather Trail Press, USA
Paris Climate Agreement (2022) UNFCCC. https://unfccc.int/process-and-meetings/the-paris-agreement/the-paris-agreement
Porritt J (2005) Capitalism as if the world matters. Earthscan, London
Princen T, Maniates M, Conca K (eds) (2002) Confronting consumption. MIT Press, Cambridge
Race to Zero (2022) UNFCCC. https://unfccc.int/climate-action/race-to-zero-campaign. Accessed 3 March 2022
Raworth K (2017) Doughnut economics: seven ways to think like a 21st-century economist. Random House Business Books, London
Regen 10 (2021) WBCSD. https://www.wbcsd.org/Programs/Food-and-Nature/News/Regen10. Accessed 15 Jan 2022
Rich N (2019) Losing earth: the decade we could have stopped climate change. Picador, London
Rio Declaration (1992) UN. https://www.un.org/en/conferences/environment/rio1992. Accessed May 2019, 5 Dec 2021
Sachs W (ed) (1992) The development dictionary: guide to knowledge as power, 5th edn. Zed Books Ltd., London
Schumacher College (2022) Dartington. https://campus.dartington.org
Sixth Assessment Report (2022) IPCC. https://ipcc.ch/assessment.report/ar/6
South African Constitution (2022) Government of South Africa. https://www.gov.za. Accessed 16 Feb 2022
Sustainable Development Goals. (2015) UN. http://www.un.org/sustainabledevelopment/sustainable-development-goals/. Accessed Dec 2015, 10 Jan 2022
UN Women (2022) UN. https://www.unwomen.org

Index

K. Blincoe, *Achieving Sustainability: The Ultimate Human Challenge*,
https://doi.org/10.1007/978-3-031-10023-9